Immortal Passage

Immortal Passage

Philosophical Speculations on Posthuman Evolution

Asher Seidel

LEXINGTON BOOKS

A division of

ROWMAN & LITTLEFIELD PUBLISHERS, INC.
Lanham • Boulder • New York • Toronto • Plymouth, UK

Published by Lexington Books
A division of Rowman & Littlefield Publishers, Inc.
A wholly owned subsidiary of The Rowman & Littlefield Publishing Group, Inc.
4501 Forbes Boulevard, Suite 200, Lanham, Maryland 20706
www.lexingtonbooks.com

Estover Road, Plymouth PL6 7PY, United Kingdom

British Library Cataloguing in Publication Information Available

Library of Congress Cataloging-in-Publication Data

Seidel, Asher, 1943–
Immortal passage : philosophical speculations on posthuman evolution / Asher Seidel.
p. cm.
Includes bibliographical references (p.) and index.
ISBN 978-0-7391-5006-1 (cloth : alk. paper) — ISBN 978-0-7391-5008-5 (electronic)
1. Artificial life 2. Immortalism. 3. Human evolution. I. Title.
BD418.8.S45 2010
128—dc22 2010028868

∞™ The paper used in this publication meets the minimum requirements of American National Standard for Information Sciences—Permanence of Paper for Printed Library Materials, ANSI/NISO Z39.48-1992.

Printed in the United States of America

Table of Contents

Preface

This book complements my earlier work, *Inhuman Thoughts: Philosophical Explorations of Posthumanity*.[1] I was then and I am now speculating on the possibility that humanity evolve to a posthuman state. As then, I present these speculations as a positive path for humanity to follow. I believe that certain deficiencies in human nature can only be remedied by a proper transformation of humans to posthumans. I further believe that a richer, more joyful existence awaits those fortunate enough to make that transition. I defend those beliefs in this book.

Immortal Passage differs from *Inhuman Thoughts* in organization, specific topics, and advocacy. I am now plainly in support of a posthuman future for humanity, whereas previously my positive attitude towards this transition was tempered by misgivings regarding what I perceived to be the escapist and utopian nature of my writings. My earlier work was largely constituted by a series of visions, among which were parallel consciousness, conceptually enriched perception, asexuality and asociality. These visions were related as aspects of posthuman existence, but they were not further organized into a developing sequence of possibilities. Rather, they were intended as on-balance, positive portrayals of such existence. In contrast, I now present a speculative movement in time, from the transitioning of humanity to posthumanity, out to roughly one million years hence. Included in this movement are visions appropriate to each temporal stage. This presentation forms a significant part of my argument for a posthuman future, an existence proof of a sketched, speculative possibility.

Although these two works have the same subject, posthumanity, they are independent of each other and they are each self-contained. The only overlap in material is the expansion of the chapter "Revolutionary Ethics" from *Inhuman Thoughts* to the chapter "The Future of Ethics" in the current work. They complement each other in that the visions of *Inhuman Thoughts* add depth to the chapter of this work titled "The Colors of Life," and the speculated historical movement within this work intimates how those visions might be extended, even as this work offers yet other visions.

As I was completing *Inhuman Thoughts* in the first years of this new century, I believed—and said in the preface to that work—that I was virtually alone in the philosophical community in my involvement with the topic of posthumanity. I knew of one other book in the area, and that book is an introductory textbook.[2] I thought I had rigorously scanned the extant literature, most of which had been written by journalists, scientific specialists, and futurists. In my defense, what is obvious now was not quite so when I finalized the version of my book that was submitted for publication. Still, there was and remains a leading voice on this topic in the philosophical community, and to my embarrassment, I had overlooked him. That is, of course, Nick Bostrom of Oxford University. I shall make such amends as I can. Bostrom's writings are significant to anyone interested in the presuppositions, commitments, and implications of posthumanism. I hope he will gather which of these writings he sees fit into a single collection. As matters stand now, he has made a number of them available at his website.[3] I urge all having interest in posthumanism to read them.

I have discovered other voices that I overlooked in my previous work. One person, James Hughes, was kind enough to disregard my lack of awareness and interview me on his radio program.[4] May every author have the good fortune to be interviewed by as kind and careful an interrogator as James Hughes. I had previously stated my belief that Francis Fukuyama's book, *Our Posthuman Future*[5] was, in my judgment, the best work I had encountered of an opinion contrary to my own. I still believe that. Allow me to add now, however, that I have found James Hughes' book, *Citizen Cyborg*,[6] to be the equal of Fukuyama's work in its rebuttal of that work.

There are two closely related associations, each affording much discussion and information regarding posthumanism and related matters. They are the Institute for Ethics and Emerging Technologies (IEET), and Humanity Plus (H+), the latter formerly known as the World Transhumanist Association (WTA). I knew of neither of these when I wrote *Inhuman Thoughts*. They both have websites[7] worth visiting.

I have received e-mail correspondence from various people interested in the subjects about which I write. Among the most prominent communications are those from Charles Tandy, whose website[8] is highly recommend to those interested in the topics discussed herein. To all others who have honored me by requesting my opinion or commenting on my work, I hereby express my gratitude.

Compared to the dating of the essays in *Inhuman Thoughts*, this book was written in a relatively compressed time period. There are advantages and disadvantages to each such mode of composition. In the case of writing this work, unlike my earlier work, I had virtually no conversation with anyone, and did not present chapters of my work at any venue, with the exception of the chapter titled "The Future of Ethics," which I discussed one evening with a group of people having philosophical interests, but who were not professional philosophers. I therefore take full credit and blame for this work.

NOTES

1. Asher Seidel, *Inhuman Thoughts: Philosophical Explorations of Posthumanity* (Lanham, Md.:Lexington Books, 2008).
2. Brian Cooney, *Posthumanity: Thinking Philosophically about the Future* (Lanham, Md.: Rowman & Littlefield, 2004).
3. http://www.nickbostrom.com/. Nick Bostrom and his colleague Julian Savulescu have recently edited a collection of readings addressing topics relevant to posthumanism: *Human Enhancement* (Oxford: Oxford University Press, 2009).
4. Interview available for listening at: http://ieet.org/index.php/IEET/more/csr20090215/
5. Francis Fukuyama, *Our Posthuman Future: Consequences of the Biotechnology Revolution* (New York: Farrar, Straus and Giroux, 2002).
6. James Hughes, *Citizen Cyborg: Why Democratic Must Respond to the Redesigned Human of the Future* (Cambridge, Mass.: Westview Press, 2004).
7. http://ieet.org/; http://humanityplus.org/
8. http://cetandy.tripod.com/cv.html

Introduction

This work is in the same general area as my previous book, *Inhuman Thoughts*. *Inhuman Thoughts* was by intention a speculative work, partially anchored in science and further informed by philosophy. I presented selected visions both of human improvement and of posthuman possibilities. I was, and am here, arguing for humanity evolving to posthumanity. To a significant extent, my argument in the previous work was constituted by an enthusiastic display of visions. There was more argument than the presentation of visions of future posthuman life, however. My overall argumentative conclusion then and now is that posthuman life, as envisioned in these books, is better than either current human life, or human life given its best possibilities as human life.

There is little said in *Inhuman Thoughts* that I want to retract here. I have expanded the chapter titled "Revolutionary Ethics" in the previous work to include an argument for transitioning that was not given there. This expanded chapter, "The Future of Ethics," aside, all else here is not a repeat or elaboration of my earlier work.

Talking to people—students, colleagues, friends, correspondents, and others—has supported what I would have supposed absent any sampling; most people believe they are happy to be human, and are not in favor of transitioning to a posthuman state, either for themselves, their descendants, or humanity in general. I have encountered people who believe differently, and as mentioned in the preface of this work, there are societies some of whose members seriously consider the idea of transitioning, and occasionally recommend it.

That some people with recognizable credentials recommend transitioning to posthumanity[1] does not appear to persuade those whose opposition I have experienced. I believe this opposition exemplifies what I consider people's psychological conatus; their tendency to desire to persevere in their given state. Of course, terminology such as "psychological conatus" is no longer scientifically acceptable. Yet the notion cuts deeper than, say, "force of habit." Most people want to persevere as who they are, and the phrase "who they are" indicates something worthy of further attention. There is also the fear of radical isolation. What if I transition to a posthuman state while those around me do not? Further, we cannot be certain that, whatever the gains in transitioning, it is not the case that more will be lost than gained. There is admittedly a problem in measuring such loss and gain, as whatever metric is considered must extend into the space of posthumanity, rather than being confined to that of humanity, the latter confine being difficult enough in this regard.

Perhaps this difficulty can be simplified by replacement. That is, rather than measuring loss and gain, those who have transitioned would be allowed to decide whether their posthuman circumstances are preferable to their former, human circumstances. There are well-known difficulties with this method of resolution, however. What if, for instance, those who have undergone some neurological procedure, which deadens a significant amount of what we intuitively consider their higher cognitive functions, report that they are more satisfied in their post-procedural state? What seems needed is a trans-invariant measure of gain, extending over the human and the transitioned posthuman. I struggle with this question further in the following chapter.

Events could determine the actual outcome of future choices regarding transitioning versus remaining human, irrespective of what people prefer today. Uncertainty reigns here. No matter what we now prefer, no matter what we might codify, nationally or internationally, unexpected events might determine, or contribute to the determination of, actions contrary to prior decisions. A wild example: extraterrestrials arrive at a crisis point. Perhaps there is a severe worldwide economic depression. Perhaps a regional conflict threatens to spread to a global conflagration, and this threat is clear and present. Perhaps there is a worldwide epidemic of unprecedented severity.

It is not difficult to imagine such crises. Nor is it unduly taxing to imagine the following. These extraterrestrials offer help. They aid our recovery from the crisis. They remain to ease our post-recovery from the crisis. We become acquainted with them through daily interaction. Perhaps there are many of them, and they manage to co-exist with us unobtrusively. Perhaps there are few of them, and the majority of humanity is acquainted with them through the media, to which the visitors allow access. It is implicit in this imagined sketch that two-way, human-extraterrestrial communication is not a problem.

Pursuing this fantasy, as time passes people come to appreciate the advanced cognitive and moral nature of the extraterrestrials. They come to wish that humanity as a whole, could more resemble these beings in their abilities and behavioral dispositions.

Bringing this imaginative exercise to a halt, I expect the following to be the anticipated point. The majority of humanity might be more willing to undergo transformation to something other than human if they experienced an appropriate, positive example of the result of transformation. This is only half the point of the fantasy, however. Let us remove much of the wildness of the preceding thought experiment. Expunge all the parts involving extraterrestrials. What remains are the crises. Suppose they are species-threatening in their intensity, whichever are occurring. Suppose further that we possess means for transforming ourselves to something other than human, something thought not to be less than human, and that such transformation gives us hope of avoiding, or at the least significantly diminishing, the crisis. I take it that these suppositions fulfilled, it is likely that much of humanity will transform.

It is also arguable that much of humanity will not transform, given the above suppositions. Successfully predicting what much of humanity will do under actual circumstances is often sufficiently difficult, to say nothing of such hypothetical circumstances.

There is an additional possibility, perhaps more likely to occasion large-scale transformation than either a worldwide crisis or a general recognition of the superiority of a posthuman existence. Discoveries are implemented. It might happen in the following manner: Major media outlets issue an initially terse report of a possible breakthrough in research on the retardation of aging. Laboratory mice somewhere have apparently ceased aging. Over some interval, say weeks, the story is elaborated; aging in laboratory mice has been reversed. Within months, results of experiments on larger mammals, dogs perhaps, have similarly indicated retardation and actual reversal of aging. By this time further details have emerged; the procedure by which age retardation and reversal is effected has to do with manipulation of material at the tips of chromosomes, somehow accomplished by pharmaceuticals.[2]

A tall story. But something somewhat in the manner of this might occur in the not too distant future. Suppose this something happens twenty years hence. What is the likely result of the general public becoming informed of this scientific breakthrough? Largely unpredictable, I venture to guess. But I believe the following is virtually certain. People will want these pharmaceuticals. Many will try to get them by any means available. There may be cautions issued by knowledgeable scientists. There will almost certainly be attempts to regulate the manufacture and release of these pharmaceuticals. On the supposition that the relevant formulas become accessible to the public, through means ranging from clandestine leaking to outright theft, and on the further supposition that these pharmaceuticals can be synthesized in moderately sized facilities, it is equally likely that regulatory efforts will be largely unsuccessful. Those who can afford the contraband pharmaceuticals will have them.

Here follows a taller story, taller than the previous tall story, I believe, because the following imagined breakthrough is less likely in the near-term. As with the other tale, this starts with a similar brief news release. In this story, the breakthrough reported has to do with cognitive enhancement—"brain boosting," if you will. By whatever means, a result has been achieved in which cognitive abilities are dramatically enhanced. This enhancement occurs across a broad spectrum of cognitive abilities. Allow me to leave the supposition in that general state. While I don't foresee quite the acquire-at-all-cost attitude among the general populace, there would be widely distributed movement towards acquisition, and as with the anti-aging pharmaceuticals, official regulation would almost certainly be ineffective, if the means were obtainable through other channels.

These two "tall stories" are given more extensive consideration in the chapter "On Transition," and located within a more comprehensive discussion of the issue of the evolution of humanity introduced here. It should be noted, however, that the large-scale realization of either of these possibilities, age retardation and reversal, and dramatic cognitive enhancement, makes for a population that has taken a genuinely evolutionary step of a sort that has not occurred for over one hundred thousand years. And, if I am correct in my supposition that many people would take this step if either of these "tall stories" were realized, then whatever current satisfaction people have with their human status is less secure than might initially be assumed.

So there are at least three reasons or circumstances that might compel humanity to transition to posthumanity in the not-so-distant future: a worldwide crisis of the sort that is perceived to be unlikely to be resolved by human means, but offers hope of resolution on the condition that humanity evolves to an indicated form of posthumanity; positive examples, present or imagined, that capture humanity's aspirations for a higher mode of existence than seems possible in human form; surprising scientific breakthroughs, such as age reversal or profound cognitive enhancement, the means of which are available to a large number of humans. These three sorts of circumstances are not exclusive, and by imaginative exercise one can effect various scenarios in which they are combined.

Inhuman Thoughts was published in the earlier part of 2008. By the later part of that year, a severe economic crisis had befallen the United States. Other countries were affected, some severely, by this crisis, which may not yet have run its course as I write these words in early 2010. This crisis has been termed the worst such since the Great Depression. That is somber news, given that some believe it took the prelude and actual occurrence of a world war to fully re-engage the productive powers of the Western democracies, powers which had been dormant or significantly reduced during the depression.

The short-sighted speculative activity that contributed to the collapse known as the Great Depression was mirrored in a speculative rampage contributing to the current economic crisis. Two circumstances are difficult to overlook. First, humanity cannot afford another global conflagration. Second, we seem not to have adequately absorbed the lesson of the previous profound economic crisis. I am not a doomsayer, and this is no warning of clear and present danger. I have no idea how matters will resolve. It may be that within a few years the present difficulty will be behind us. It may even come to pass that we implement prudent and long-lasting controls on speculative activity, and that we thereby successfully prevent further economic disasters. I am not competent in these matters, and there are few willing to place full confidence in the long views of those with ostensibly proper credentials.

The above should serve as a reminder of the precariousness of our (that is, humanity's) situation. Others call attention to other dangers. As a species we have faced many catastrophes and obviously survived them all. While I would not consider our survival to date to furnish the strongest inductive evidence of our continuing ability to survive, I believe it gives worthy evidence of the resilience and resourcefulness of our species.

These various crises have been overcome by us as a species, but they have almost invariably been attended by much suffering. Thinking in terms of the overall welfare of our species, both at some historical moment and over time, perhaps it would be better for us were we able to avoid much of this suffering. If this last thought seems absurdly obvious, it is well to recall that many believe we have become better, as a species, through our overcoming of these crises. We have been driven to innovate, to form cooperative unions, to pursue peaceful resolutions. Further, it is often said that the human qualities that contribute to some of these crises, greed for instance, also contribute to positive aspects of our growth, individually and collectively, so that radical alteration of such characteristics might well produce unanticipated overall negative consequences.[3]

I do not share the thought that human nature is better left undisturbed. Rather, I believe that we carry characteristics some of which have outgrown their survival value. I believe further that our increasing scientific and technological sophistication, wedded to these outdated characteristics, is at best a troublesome union, at worst species-threatening. Still further, I believe there is a timetable for us to alter ourselves; that is, to alter our biological, psychological, and sociological nature, so as to avoid the increasingly dangerous potential consequences of our nature. I am not a doomsayer because I believe such alteration is possible. I am not issuing a warning because I realize I could well be mistaken in these beliefs. I also believe there are more positive reasons for altering our nature, aside from crises-avoidance. Still, the attitude manifested in this introduction, especially in this paragraph, contrasts with that manifested in the beginning chapter of *Inhuman Thoughts*. There I was apologetic. I considered the challenge that my work was bourgeois, utopian, escapist. I acknowledged that there was some truth to those criticisms. I still acknowledge this, but to a lesser extent. Although I would rather emphasize the positive reasons for transitioning to posthumanity, and this is what I am for the most part doing in what follows, I have come to believe that there is some notable evidence supporting the choice to transition to avoid various crises. If I am correct in my assessment of the possibility of these crises, then perhaps my appeal is not as utopian as I earlier thought.

Philosophers typically argue for their points. My overall point, as I have stated, is that humanity should evolve to posthumanity. By themselves, visions are not arguments, and as with *Inhuman Thoughts*, so is this substantially a book of visions. Visions can serve argumentative purposes, however, within an overall structure. My earlier work has loosely connected visions, which are intended to show the possibilities of superior lives attending certain forms of posthuman existence. These visions are subjected in that work to critical philosophical scrutiny, rather than unquestioning enthusiasm. Still, they are offered as illustrations of possible aspects of posthuman life, the better to praise such life.

Philosophical argumentation by means of visions has a history. Perhaps Nietzsche's *Zarathustra* is too extreme an example for some, but there are other examples more generally accepted as constituting a philosophical argument. The most familiar such instance for readers of this work is likely Plato's *Republic*. Having dialectically dispatched Thracymachus in Book I, Socrates is challenged by Glaucon and Adeimantus to give a more convincing account of the just life. The resultant account is primarily of the just society, in which just individuals may flourish. This visionary account is delivered at length. Viewed as an argument, the movement from Book I through Book VII may be summarized as follows:

1. There exists a prima facie plausible conceptual scheme involving ruling (including self-ruling).
2. In this scheme, the notion of ruling is connected to the notion of benefit (good), and the manner of connection is that rulers rule for the benefit of the ruled.
3. People are self-ruled insofar as their self is optimally ordered.

4. People are most likely to achieve self-rule in a society that is appropriately ordered.
5. Achieving such a social organization requires a radical re-organization of present society.
6. Effecting and maintaining such re-organization requires appropriate knowledge (ultimately, knowledge of "The Good").
7. Hence, to instantiate a society whose members maximize their benefit, society must be reorganized to properly educate and maintain rulers, who will apply their knowledge to create and maintain those conditions offering the maximum number of society's members the maximum benefit available to them.

This is of course only one of various ways to summarize quickly the movement from Book I through Book VII. It should be apparent that what Plato terms "justice in the individual" is what I am terming "self-rule," as the latter is a more convenient term for my purpose. I take Plato's at-length argument to be in response to the claim that people are only apparently willing and capable of self-rule. Rather, so Plato is arguing, there are circumstances which can be envisioned in which people in fact achieve self-rule, either under the direct guidance of appropriate knowledge, or under the guidance of those having such knowledge. Since Plato's Socrates does not claim to have such knowledge, the scheme of the *Republic* is hypothetical. Hence, Plato's argument, as I have summarized it, is a "best that one can do" sort of visionary response to the challenges presented at the beginning of Book II.

Not everyone acquainted with Plato's thoughts admires them. A reading of *Inhuman Thoughts* reveals that I do, although from my historical position I have reservations about the proposed scheme of the *Republic*, as noted in my earlier work. I have no reservations about the general argumentative structure of Books I–VII of the *Republic*, however. Plato knows what he is up against, as put in the mouths of Thracymachus, Glaucon, and Adeimantus. Most important, Plato is aware that the points they make are unanswerable within present society. I largely agree with Plato regarding this, but I go further in arguing that those points are unanswerable within the present confines of human nature. The phrase "human nature" is of course philosophically problematic, but I believe on faith we can acquire a more scientific understanding of ourselves. In the following chapter, "The Future of Ethics," I elaborate on these points.

Like Plato, I try to sketch a positive vision of transformed, sentient, cognitive life. Unlike Plato, I do not think it will be a human life. The presentation that constitutes the majority of this work is hypothetical. I do not believe it is a likely story. I believe it is a possible story. I intend that the presentation afford a glimpse of the possibilities of posthuman life, taken as a progression from the present to what I term "the deep future." I present this progression in the spirit in which Plato delivered his scheme; I want people to respond positively to it. Challenged to show that posthuman life is the more desirable alternative for humans, rather than that of remaining human, this is my response.

Prior to my extended discussion of transitioning to posthuman life, I establish some ethical, metaphysical, and epistemological requisites. Ethical requisites are considered in "The Future of Ethics." Within the chapter titled "Minds and Related Matters," I stipulate several metaphysical and epistemological conditions that should carry through both the transitional period, and at least the first phases of posthumanity. These include genuine consciousness, a realistic attitude towards the external world, and an intuitive continuity of selfhood similar to that which we ascribe to human life.

My discussion of the movement from the human to the posthuman begins in the chapter "On Transition." I first elaborate on the relevant variables constituting an idealized transitioning; one that occurs with a minimum of discomfiture on the part of an informed worldwide human population. I then consider more likely scenarios, none of which are fully satisfactory from the standpoint of the best possible transitioning. These less-than-ideal cases involve careless exploitations of technological breakthroughs, as well as uncertain relations between those who have transitioned and those remaining fully human. My expressed hope is that virtually all humanity will eventually transition to entities having greatly extended lifespans and considerably enhanced cognitive powers. Having those enhanced cognitive powers, the worldwide population of evolved humans (transhumans or posthumans) will hopefully have gained the requisite understanding to live together peacefully and supportively.

The three chapters following the chapter "On Transition," titled "The Colors of Life," "The Far Future," and "At the Limits of the Conceivable," sketch a speculated temporal progression of posthumanity. "The Colors of Life" considers posthuman life to roughly ten thousand years after the transition. "The Far Future" speculates on the interval from approximately ten thousand years out to one hundred thousand years. "At the Limits of the Conceivable" ranges over what I term the "deep future," the period from the far future out to approximately one million years.

Although these chapters are the core of this work, they should be approached with a certain lightheartedness. I of course do not have the slightest clear picture of what will prevail one million years hence. This is speculation which, if offered as serious prediction, would outdo anything in degree of fantasy but some science fiction and some theological prophecies.

Such speculation belongs in an ostensibly philosophical treatise only to the extent that it serves some philosophical purpose. I claim two philosophically related results from this deep future speculation. For one, it is addressed to the criticism that, as posthumans living indefinitely extended lives, there is no manner in which it can be considered that these lives are neither indolent, nor in some other way less desirable than human lives of limited duration. Second, this speculation indicates, albeit sketchily, manners in which these sentient beings might overcome various epistemological and ontological limitations.

These three chapters share the general characterization of being futurist speculation. Nevertheless, there are philosophical discussions particular to each of them. "The Colors of Life" defends the claim that, on reflection, most humans value their cognitive mind above all else. Establishing this claim is important within the context of this work. In an earlier chapter, "The Future of Ethics," I introduce the notion of "pointers" that allow for a bridging of normative judgments from the human to the first phases of the posthuman. The argument in support of valuing our cognitive powers above all else lends some closure to the earlier discussion of pointers.

The chapter titled "The Far Future" includes a discussion of the claimed necessity of pain and suffering. "Necessity" is used here in the sense of "requisite for survival." I also consider the claim that humans (or human-like beings) are on the whole benefitted by overcoming pain and suffering. There is additionally some consideration of self-identity over time, given the speculative possibilities regarding the manner of our far future descendants.

As these speculations project further into the future, they are less detailed, more general. In "At the Limits of the Conceivable," speculation regarding our deep future descendants (or descendant) extends to questions of alternative universes and the status of logical and mathematical truths. A remarkable possibility is the power of these speculated deep future descendants to create universes, wherein our present understandings of what is conceivable might be altered.

The final chapter of this work, "Loose Ends and Final Thoughts," is a deceleration from the speculations of the preceding chapters, and a return to considerations of normative ethics. In particular, I discuss various chronic problems that confront us today, with an eye towards their resolution in our transitioning to a posthuman future. I conclude with a defense of the thesis that normative disagreements are disagreements regarding facts. As a final note, I observe that if this thesis is correct, it behooves us to optimize our ability to entertain relevant facts, which is best done from a position of appropriate posthuman evolvement.

NOTES

1. Some such people are noted in the preface of this work. See also various chapter notes and the bibliography.

2. This speculation has some factual basis. An optimistic case for the scenario I imagine can be found in Aubrey De Grey and Michael Rae, *Ending Aging: The Rejuvenation Breakthroughs That Could Reverse Human Aging in Our Lifetime* (New York: St. Martin's Press, 2007).

3. Francis Fukuyama makes a similar point regarding general human characteristics in *Our Posthuman Future*.

Chapter 1

The Future of Ethics

In this chapter I want to advance the radical claims that the future of ethics should and likely will depart from its past and present in a significant manner.

There are obvious ways this has already happened. Neither Plato nor Kant could have contemplated many of the questions concerning bioethics, for example, although their ethical theories are applicable to at least some of these questions, if not all of them. What I shall say departs from the example of bioethics in that it is broader, and more far-reaching. I shall maintain that the study of ethics has not yet realized its implicit promise, and that it cannot realize this promise, given human nature as it is presently constituted.

These statements are made with the understanding that a significant task of ethics is the delivery of a theory of how to live, and that such a theory must take account of human nature. Were it not for certain thinkers, perhaps most prominently Sartre, the wedding of ethical theory with human nature would be unproblematic. I mention such dissent as that of Sartre only to note it. I shall not attempt to controvert Sartre's claim that human nature is chosen, rather than given. I don't believe this claim to be in the main true, but anyone who does will not accept what follows.

I shall argue that human nature is faulty to the extent that human nature, as it currently exists, does not allow a satisfactory implementation of the sort of practices an as-yet-undiscovered successful ethical theory would demand. As a consequence, human nature must be altered if humans, or posthumans, are to accommodate the prescriptions of such a theory. The future of ethics, I shall urge, lies in the co-development of this theory together with those future scientific theories comprehending the possibilities of an evolved human nature.[1]

Hopefully, calling for an altered human nature will not seem thoroughly outlandish. Such a call has a long history in philosophy. When Socrates urges his followers to "practice death" in the *Phaedo* (61–66), he can be interpreted as recommending an altered form of life in which the needs and cares of the body are minimized in favor of activity of the mind. His call for such practice, considered in the context of his extended polemic against the body, is considerably more radical than, say, Mill's assertion that the life of intellectual pleasures is better than that of "lower" satisfactions. Then there is Nietzsche's Zarathustrian vision of the overman, a vision at considerable remove from the Platonic ideal. Still, Zarathustra's words that man must be overcome have a general commonality with those uttered by the Socrates of the *Phaedo*. Historically situated between Plato and Nietzsche, Spinoza offers an ethical theory for the few, lamenting that the many will not overcome their nature so as to lead the best lives. And in the twentieth century various revolutionary political leaders have sought a transformation of "alienated man" to "socialistic man." Such political movements have intellectual roots in well-known philosophical writings.

There are varying expectations of useful results in the various philosophical fields. Those philosophers centered in metaphysics, for example, might endorse nominalism or realism with some fervor. Typically, those whose philosophical inclinations lie outside such debates have relaxed attitudes toward their outcome. Epistemological concerns, such as foundationalism versus coherentism, or internalism versus externalism, similarly proceed seemingly interminably, of strong interest only to their limited audiences.

Such mildly disdainful remarks invite response. Of course esoteric subjects are of interest to a comparatively tiny community. Even within a limited overall discipline such as philosophy, specialization often precludes ecumenism. On rare occasions, a startling approach, a novel theory, attracts common attention. Such moments are exceptional, even though their collective happening over historical intervals is largely definitive of the general field. What more should one expect from a humanistic study?

This response is not altogether satisfactory in the case of ethics. Granted that various philosophers discuss such topics as varieties of utilitarianism, and deontological versus teleological moral theories, with all the involvement of those engaged in the above-mentioned metaphysical and epistemological disputes. Let it be granted also that such discussions have similar limited appeal. Still, the general philosophical audience (larger than the circle of professional philosophers, some of whom are at times not part of this audience due to their professionally limited concerns) expects more from ethics. Occasionally, their expectations appear to be met. The excitement attending the pronouncements of existentialist thought in the immediate postwar years exemplifies both this wider yearning and its impermanent satisfaction. As with other such moments, the uptake of existentialism was temporary, for the reason to be given below.

Outsiders expect little from metaphysics, primarily because natural science, particularly physics, has addressed the overall question of the nature of reality in what many regard to be a more satisfactory manner. In particular, practitioners of the various natural sciences are seen to be more in agreement than metaphysicians regarding the "deeper" facts about reality. Further, natural scientists amaze and delight us with their explanations and predictions, and their findings are often of use to us in our daily lives.

Although once again, the intellectually alive segment of the general public has little to do with the various disputations of epistemology, there is not the analogous satisfaction, as in the case of metaphysics, in the deliverances of natural science. Instead, there is an awareness of various movements under the general heading of "cognitive science," such as cognitive psychology, decision theory, artificial intelligence research, and other related fields. Epistemologists have taken some note of these trends through the relatively new sub-fields of naturalized and social epistemology. Still, it is the cognitive scientific fields that draw more public attention, and such prevalence is arguably due to various expectations that research in these areas will produce explanatory and practical fruit.

These remarks regarding metaphysical and epistemological concerns are likely to annoy. They are overly broad and they are, at least to some extent, controvertible. I can offer commiseration. Before this study terminates, I will be in a similar if not worse position to that of the dueling metaphysicians and epistemologists. One need only revisit my opening statements to appreciate this. Much of what I will urge, to the extent that it is read with the seriousness I intend, will be taken to be ludicrous or outrageous. Inasmuch as I shall argue that humans ought to cease being humans, the term "repugnant" will likely come to some minds, all the more as I suggest manners of posthumanity.

In a society productive of sufficient surplus value and tolerant of varying viewpoints, philosophical speculation and disputation are and ought to be permitted. Perhaps one also accepts the facile dismissiveness of those thinking such activity is and ought to be relegated to its practitioners, and to such café intelligentsia as follow it. Thoughtful members of the general public have a prima facie right to request something more from the field of ethics, however, given the ostensible relevance of ethics to daily lives. Practitioners of ethics might be said to owe society a return on its investment, such return taking the form of findings having practical implications.

A loose scrutiny of twentieth-century English-speaking philosophy, so-called "analytic" philosophy, leads one to suppose that this group of philosophers has not given society the normative ethics that might be considered society's rightful claim. A closer look suggests a modification of such judgment is in order. Perhaps most significantly, the field of applied ethics—business ethics, bioethics, environmental ethics, to name some of its sub-fields—has been offering normative thoughts for the past several decades. Whether most of the literature of applied ethics lies within the category of analytic philosophy is debatable, but what is not debatable are the practical concerns applied ethics pursues.

Turning to the largely non-English-speaking continental European philosophy, one finds much normative ethics. Although one might question whether the utterances of the practitioners within this field have practical value, it is incontestable that much recent continental philosophical thought has influenced the lifestyle of many, including a sizeable number of non-professional philosophers. To this extent, at the least, such thought has had value-in-practice.

It would seem then that applied ethics and continental philosophy, their marked differences notwithstanding, have been fulfilling the promise of ethics—to supply a theory of how we ought to live. Still, one might raise doubts. Applied philosophy has addressed many questions pertinent to practices in roughly circumscribed domains. Yet whether taken individually, as bioethics, business ethics, environmental ethics and so on, or collectively as the sum of all thought in these fields, much of normative importance remains outside the discussion within these fields. The question of how to live one's life is only partially addressed by applied ethics, to the extent that matters in one's life fall within the appropriate areas of concern to be deemed matters addressed by applied ethics.

Much continental philosophy aggressively addresses the overall question of the manner in which to live one's life. Yet there is an apparent self-centeredness to much of the normative discussion in continental thought. This impression has minimally the following related sources: the emphasis on subjectivity, the appeal to a "will to power," and the concern with "the other" as the not-me who impinges on me.

I do not want to maintain that the above is a comprehensive, or even a thoroughly fair, summary of the current situation in ethical thought. There is a strong current of socio-political focus in normative continental philosophy, drawing from sources such as Hegel and Marx, and exemplified in the writings of the Frankfurt school, for example. Much analytic philosophy is done at article length, rather than book length. It is therefore possible that a proper selection of such analytic-philosophical essays, suitably conjoined, instantiates a coherent and full-range ethical theory.

The preceding charitable response to the challenge that the practice of ethics has not succeeded in what the thoughtful general public regard as its mission, will be argued to fail. Prior to stating the argument, however, I want to examine an ethical theory that I consider to have taken a correct approach to the question of the manner in which to live one's life, albeit ultimately unsuccessful in delivering a satisfactory answer to the question.

The theory I have in mind is that of Plato. However, it is less than granted that Plato has one such theory, rather than several, or that if one, that it is self-consistent. These are legitimate questions which I shall not pursue. Instead, I am interested in what I take to be Plato's approach to the task of developing a successful ethics.

Early in the *Protagoras* (329ce) Socrates raises subtle questions regarding the unity of the excellences (i.e., virtues); whether the unity is as parts of a face to the whole face or as the smaller golden parts of a larger lump of gold. This question of the overall unity of the five excellences is pursued with logical sophistication throughout the dialogue. In various dialogues Plato offers altered definitions of familiar normative notions such as courage (*Protagoras* 360), justice (*Republic* Bk IV), punishment (*Gorgias* 478), and ruling (*Republic* Bk I). While the alterations themselves are interesting, of further note is their relation to their more prosaic counterpart definitions. In accordance with Plato's belief that the world of experience receives such intelligibility as it has from its "participation" in an integrated system of "forms," and that ordinary thought and language offer initial indications of this intelligibility, although in a dim and confused manner, Plato seeks what will in later times be termed a "conceptual scheme" having both a large degree of internal coherence and a certain sort of applicability to the world of experience. In particular, this scheme applies to the perceived world in the manner that it is most good; that is, it is of most benefit to us to order the world according to that scheme, both in our knowing the world and in our practical activities.

Not all historically prominent ethical theories have proceeded in the manner ascribed to Plato. Some of the most famous such theories have sought to refine our common understandings of ethical terms, without straying from our ordinary intuitions. The resulting coherent theories are hence commonsensical. Mill's *Utilitarianism* and Aristotle's *Nicomachean Ethics* are separate versions of such an endeavor. Other well-known theories have aspects of Plato's approach. Kant, for example, refines and systematizes the notion of morality in a manner that had not occurred previously. However Kant, unlike Plato, is not interested in integrating moral and nonmoral goodness. Involvement with the latter, Kant asserts, leads to a "heteronomy," a relativism of valuation lacking the unqualifiedness of the lawful nature he believes to be possessed by a genuine morality. Whereas Plato is interested in both moral and nonmoral value (although lacking the term "moral"), he clearly preferences benefit, or nonmoral goodness, choosing to argue that such benefit necessitates a virtuous life. In this regard Spinoza's method is closer to Plato's, inasmuch as his technical system of definitions both encompasses personal benefit, which is primary for Spinoza as it is for Plato, as well as being other-regarding in the interest of personal benefit, as it is for Plato. Still, Plato sets the goal of such endeavor (at least in the *Republic*) as maximizing the well-being of all, or as nearly all as possible, whereas Spinoza despaired of the many following the path to "freedom." Admittedly, Plato's idealized republic has a rigid class structure with different lives for those falling into the separate classes, but perhaps if he could have envisioned the productive possibilities of an industrialized society he would have made allowances for the many to pursue the life of mind.

There are of course many other ethical theories. I cannot investigate them all. Nor can I claim to have given anything approximating a comprehensive examination of those I have singled out. Yet I am confident that ethics, past and present, Western and Eastern, analytic and continental, have failed to deliver a successful theory of how to live. There have been ethical systems, or systems of thought from which a coherent ethics can be obtained, having had profound influence on human affairs. Marxism and classical liberalism are two examples of such. Yet these systems have had less than universal acceptance, and societies that have been at the least nominally guided by them have not fulfilled another important criterion for a successful ethics.

In the preceding paragraph I have indicated two necessary conditions of a successful ethical theory; universal acceptance and something further that must be possessed or exhibited by those accepting such a theory. Unless more is said, such conditions are vague to a degree approaching vacuity. Is universal acceptance temporalized? That is, need this condition apply for a minimal epoch? Need it be literally universal, or near-universal? Since infants and psychotics cannot be expected to entertain an ethical theory, if such human beings are counted among the universal population, obviously near-universal will need to suffice. But then, what is intended by "acceptance" of such a theory? Does it require complete understanding (including all implications)? And so forth.

These difficult questions arise in consideration of the specified condition. The other condition of a successful ethical theory is that it be a significant factor in enabling us to achieve, or approximate, a near-ideal state of affairs in which the overwhelming majority of people live satisfied lives at their highest potential, such potential being characterized socially, psychologically, and materially.

The difficulties besetting the criterion of universal acceptance appear comparatively less than those attending this seemingly utopian condition. What constitutes satisfaction, and in what manner can the highest potential of a life be non-controversially determined? Characterizing highest potential in the three dimensions of the social, the psychological, and the material does not reduce the difficulties. Does the material dimension include physical health, for example, and how is such to be specified—there being conflicting criteria even in this limited case. Then there are the potential inter-dimensional conflicts. It is not settled that psychological goals of the highest potential are fully consistent with what might be specified as social goals of the highest potential.

These critical comments, and others left unsaid, lead one to suppose that any attempt to characterize the success of an ethical theory along the lines proposed is desperate, if not hopeless. Yet I am not entirely convinced of this melancholy result. I see an improved possibility of a successful ethics, characterized in the above terms, where such improved possibility is based upon two considerations. The first consideration is that the criteria of near-universal acceptance and that of a near-ideal state of affairs might well be mutually supportive, in that recognition of the sort of state of affairs that would constitute such an ideal would tend to induce near-universal acceptance. The second consideration involves a comparison with major events in the history of science.

Thomas Kuhn's account of scientific revolutions[2] can be, and has been, challenged on some of its major points. These challenges accepted for the purpose of this discussion, there remains the following point of emphasis. Kuhn reminds us of what the circumstances were before and after various major breakthroughs. Such breakthroughs often did not simply amount to the discovery of significant new regularities. Rather, they involved the reformulation of basic concepts, together with the establishment of concomitant and subsequent mathematical and physical techniques. As Kuhn emphasizes, they also involved the perception of practitioners that these breakthroughs (or "paradigms") had gotten matters right. Some outstanding examples of such breakthroughs (not all of which are Kuhn's examples) include Newtonian mechanics, Daltonian chemistry, and more recently, plate tectonics.

Plato's various redefinitions of familiar normative notions, noted above, have some resemblance to Newton's careful construction of a technical system of definitions in his *Principia*. As was Plato, Newton is seeking an integrated system of concepts which have their origins in vague and confused ordinary thought and language correlates. The point of Newton's system is to achieve an explanatory, predictive, and manipulative grasp of the reality constituting its domain. So with Plato. Unlike Plato, however, Newton's system has had near-universal acceptance among the world-wide educated class. This comparative failure of Plato is understandable not only in terms of the different domains to which their inquiries were directed. It should be kept in mind that Plato and his contemporaries fared little better in delivering successful natural scientific theories. Qualitatively and quantitatively better data and better instrumentation, among other requisites, were lacking in Plato's time.

It remains that no ethical theory has approached the near-universal acceptance of Newtonian mechanics and other similarly well-regarded natural scientific theories. My diagnosis of this situation is twofold. First, there is the lack of near-universal agreement among the educated class that any one theory has gotten matters right in the domain of normative concern, in anything like the manner of the agreement that various natural scientific theories have gotten matters right. This agreement could be secured rather quickly, given the current worldwide situation of information propagation. Allowing that the propertied class manages a significant amount of the informational media and generally acts in its interest, the various modes of interconnectedness, from physical travel to internet messaging, circumvent any such complete control by vested interests. Humanity has now, perhaps more than at any previous moment, realized Mill's "marketplace of ideas."

Because I want to emphasize my second reason for the lack of near-universal acceptance of any ethical theory, I find it useful first to examine other reasons given for what I consider the lack of a successful ethical theory. In no particular order, here are some of the most significant such reasons:

1. No such theory is possible because humans are radically free to choose any set of values.

   This position is most closely associated with the earlier views of J. P. Sartre.[3] While I do not believe Sartre's position is correct, I cannot address it adequately here. If it is correct, my outlook is mistaken.
2. No such theory is possible because normative terms are non-referring, and hence normative statements have no truth-value.

   This is of course meta-ethical noncognitivism. I do not accept this theory, but if it is correct then my outlook is mistaken.
3. On the assumption that a successful ethical theory is partially founded on an adequate theory of human nature, no successful ethical theory is possible because no adequate theory of human nature can be specified in practice, although such a theory is in principle possible.

   What is being argued here is that, unlike lifeless, idealized point-masses, human beings are far too complicated in their psychological and sociological aspects to be the subject of lawful generalizations similar to those in the natural sciences. Hence, in speaking of the non-biological characteristics of human nature the most that is available to us are loose, statistical

> generalizations. Further, there is disagreement as to the appropriate level at which to specify such generalizations, as well as disagreement as to which set of concepts are proper to organize such generalizations.

This last seems a fair statement, if obviously overly broad, of the state-of-the-art regarding successful theories of human nature. Practitioners and exponents of particular psychological and sociological theories purporting to be overall theories of human nature will not readily accept my fair-statement judgment, but neither do proponents of particular theories of normative ethics acquiesce to the judgment that such theories are unsuccessful. Yet I remain unconvinced that any conclusion follows, or is even rendered probable, regarding far-term scientific theorizing on human nature. What over-the-horizon scientific theories, such as those directed towards matters of the brain, might discover, and what impact such discoveries might have on overall theories of human nature, seem to me not decidable on the basis of present practices.

In considering the last of these three reasons, I have implicitly accepted the assumption that a successful ethical theory will rely on a successful scientific theory of the psychological and sociological aspects of human nature. There is the suggestion here that such an ethical theory will be of a scientific nature, rather than mainly of a philosophical nature. I do not wish to withdraw that suggestion, nor do I wish to argue in support of it. It seems to me that the only theories that have the near-universal acceptance of the various showcase examples of natural science are themselves other scientific theories. It also seems to me that metaphysical speculations on the nature of causality, for instance, proceed in the usual pro-and-contra-manner of philosophical disputation—which proceeding is acceptable to the properly educated only because natural science has progressed to the extent it has. Put another way, held against the successes and implementations of natural science, philosophical speculations—mine here included—are typically luxury items. I do not mean to suggest, however, that there is a clear separation between philosophical matters and natural scientific matters, nor that such scientific matters are never the subject of philosophical debate, even among scientific practitioners, nor that scientific matters, especially frontier scientific matters, are not subject to disputation among practitioners. I especially do not mean to suggest either that all philosophical efforts lack profound utility, or that philosophical thought cannot serve in the development of scientific thought.

The foregoing remarks indicate my acceptance of the difficulty of arriving at a successful ethical theory, based at least in part on the complexity of human nature. If I thought that our current lack of scientific understanding of this complexity was the major obstacle preventing the discovery of a successful ethical theory, the remainder of this study would focus on speculation regarding the possible paths leading to such an understanding of human nature. Unfortunately, as tenuous as such speculation would be, I fear that what I am about to say regarding my second reason for the lack of near-universal acceptance of an ethical theory will appear to make matters worse. In brief: I believe that what we currently understand about human nature is sufficient to show that a successful ethics is not possible, given human nature as presently constituted.

There is some reason to be hopeful. Assume science and philosophy can co-develop a theory, or perhaps a spread of theories, indicating in what manner humanity should evolve so as to optimize a form, or forms, of life. Suppose such optimization is grounded in near-universal acceptance from the present standpoint of humanity as homo sapiens, as well as from the speculated future standpoint of humanity (or posthumanity). These suppositions fulfilled, we will possess the ethical theory we have been seeking. Indeed, such a theory, together with the good life it details, might prove more accessible if as a species we undergo an appropriate (by the light of that theory), controlled evolution.

But why suppose that our current nature, by which I mean our physiological and psychological constitution, does not allow a successful ethics? Put directly, I believe there are irreconcilable conflicts arising from human nature as it is. I shall give two examples of such conflicts, one of which may appear frivolous—until one entertains it thoughtfully, the other almost certainly will be taken seriously, although some will doubtless believe a response is possible.

The age-old conflict of freewill and determinism provides an example of a possibly irreconcilable conflict within human nature. In the recent philosophical literature on this topic, there has been much focused, careful discussion.[4] Rather than delve into this large area of study, let us make the following suppositions: compatibilist resolutions of the conflict are unacceptable; contra-causal behavior is either conceptually incoherent, or not present in nature; to the extent that human behavior is ontologically indeterminate, it is not ascribable to human agency. Not everyone involved with the freewill-determinism discussion will agree with these assumptions. But suppose future scientific discoveries banish libertarian possibilities. Suppose that compatibilism is either rejected, or allowed to be a fair account of much of our ordinary outlook on behavior, but not applicable as a resolution of the metaphysical problem. Finally, suppose it is granted that, in the "deep, metaphysical sense of the problem," human behavior is not free.

Much of our outlook is retributivist when it comes to assessing behavior that we find disagreeable. We seek payback, as individuals and as a society.

In the case of relations between individuals, such behavior is often immediate. In the case of our social arrangements we institutionalize such phrases as "the punishment should fit the crime," meaning the wrongdoer should be made to suffer to the extent that the wrongdoer's action occasioned suffering. Granted that not everyone reacts to instances of behavior of others that are typically reaction-triggering. Granted also that not every legal practitioner or theorist is a retributivist. Still, these observations are accurate for the overwhelming majority of individuals and societies.[5]

Not too long ago educated people lived their lives aware that many of their fellow human beings were bought and sold, enslaved. Often enough, these educated people believed that those enslaved were of the same species, with the same potential, as their own children. Yet for a period of generations, they managed to live with whatever qualms they had, until such time as the situation became sufficiently intolerable for them. Can we not expect similar cognitive dissonance should science effectively demonstrate the lack of libertarian free will in humans? If so, what options are open to us?

Suppose, as might well be the case, our retributivist dispositions are too deeply entrenched in our psychological nature to yield to education, or other forms of social conditioning. If we cannot accommodate this contradiction between what we know and our manner of behavior, we may arrive at a crisis resolvable, if at all, through controlled evolution to something other than what we are now. If, that is, the assumed circumstances are realized, we may have no problem recognizing the immorality of reactivity and retribution, but such recognition will make it all the more difficult for us to live satisfied lives if our present nature prohibits our behaving in any other manner towards others. Granted, we can alter our criminal justice system so as to banish retributive punishment. As presently constituted, however, we might not be able to alter our daily attitude towards others in a similarly rational manner.

The number and manner of suppositions in the above discussion of the conflict between retributivist attitudes and hard determinism are such that many will remain unconvinced that this speculated conflict exemplifies an irreconcilable difficulty, given human nature as presently constituted. I believe that even if we humans possess deep-metaphysical, libertarian, freewill, such freewill manifests itself so infrequently that we ought not hold the attitudes we hold towards one another. I believe that ridding ourselves of these attitudes requires profound alterations of the nature that has served, but outlived, its evolutionary purpose. However, I cannot defend these beliefs here.

There are other irreconcilable aspects of our nature whose telling arguably requires less defense. To cite one such aspect, consider the tension between our beliefs regarding equality and our motivation for self-enhancement. Liberal democracies typically embrace equality as both a foundational requirement of their constitutions and as an ideal whose further implementation is among their foremost concerns. Waiving the well-known ambiguities of "equality," we need only focus on the tension between classical and contemporary liberalism. The former emphasized freedom of individuals to pursue their self-betterment, the latter seeks a leveling of the disparities of wealth which manifest themselves as various inequalities. Many of us know that economic disparities are often inherited, and many of us believe these disparities are significant factors in the dispirited lives frequently associated with those who are comparatively impoverished. Many of us wish for a world in which, whatever cultural differences are manifested among the educated classes (assuming an intuitive, standardized notion of "educated" is possible), a maximum of the world's population live the lives currently enjoyed by members of the educated class.

The sort of equality considered here is to some extent opposed by the popular attitude of "enjoying the fruits of one's labor." The spirit of competitiveness allied with this attitude is justified, many believe, by the gifts brought about through innovation and enterprise, such gifts being widely disseminated. Elaborate justifications have been offered for the competitiveness ostensibly in conflict with the ideal of equal distribution of goods and services. Others have been less sanguine about this conflict. Freud despaired of a benevolence unfettered by the aggressiveness he regarded as an innate part of our psychological constitution.[6] All parties to the dispute regarding the value of the conflicting human tendencies of wishes for a society of equals, and drives for self-betterment, agree that these tendencies are prevalent in many individuals and that they are in conflict. Some disputants will argue that one or the other of these tendencies is learned, the other being innate. To date, however, there exists no theory, acceptable in the manner in which various natural scientific theories are acceptable, offering an account of the manner in which either of these tendencies are generally removable, while preserving other aspects of human nature commonly thought worth preserving.

This is not to say that a complete removal of either of these tendencies is desirable. We have no theory that would decide in favor of such removal. What is being noted is the absence of the means of thoroughgoing control of such characteristics, should such control be deemed appropriate. In general, there is a lack of theoretical understanding of the social role of various characteristics distributed within a large population. There may be proposed answers to questions such as, "What would be the effect of a general lowering of aggressive behavior in a large population?" But such proposed answers do not proceed from any theory with a degree of acceptance approximating those of well-confirmed natural scientific theories.

At minimum, what is needed to answer the sort of questions posed above is a comprehensive theory of human psychology. Despite the likely objections of those adherents to a particular approach, such as behaviorism or neurophysiology, there is no such theory. There is no near-universal agreement among practitioners as to what level or levels such a theory should occupy (assuming agreement as to what constitutes these levels, which is problematic), nor is there agreement as to appropriate concepts within the higher levels, nor are there what in Kuhnian terminology would be understood to be "paradigmatic moments" unifying practitioners. Of course, if thinkers such as Sartre are correct, there could be no higher-level science of human psychology that would be definitive of human nature.

Against the backdrop of these observations, one can either be silent or speculate. I choose the latter option, starting with the assumption that Sartre is incorrect in that there is a discoverable, although not yet discovered, correct and comprehensive theory of human nature in its psychological and sociological aspects. As I have suggested above, I believe that such a theory would reveal irreconcilable conflicts in human nature, preventing a coherent ethical theory maximizing the best life for the largest number of people. I do not believe that these conflicts can be overcome, or sufficiently mitigated, by educational reforms, pharmaceuticals, minimal genetic engineering, or any combination of such, including related approaches. I therefore choose to speculate regarding more profound changes to human nature in the sometime future.

Advances in knowledge might open possibilities. There are constituents of human nature, as such is commonly understood today, that in the sometime-future might be considered to have outlasted their evolutionary role in human survival and prosperity. As an example, consider our procreative desires, evolutionarily programmed as sexual desire/gratification. If there is a species-wide possibility of indefinitely extended life, and if this is conferred on the many rather than the few, then there will likely be a limitation on the possibility of future generations. Of course, we could continue our sexual behavior without the possibility of children, but unfulfilled of its procreative function such behavior might, in the passage of time, seem a holdover from a more primitive past. There are also the various conflicts that arise from mating practices, such conflicts being avoidable in the absence of these practices.

Within the preceding paragraph I have suggested, as a positive future for humankind, indefinitely extended life, childlessness, and sexlessness. Should such a vision not be sufficiently adverse, let me add the possibility that in this future we not have need of one another's social company. This addition is not presented in the spirit of disputation, but rather to remove a further source of conflict among our speculated descendants.

Such speculated descendants will thus have natures—biological, psychological, and sociological—at some remove from our own. If these or other speculated alterations in current human nature are granted some plausibility, then as means of securing such alterations come into view a significant task of ethics will be that of evaluating competing visions of an evolved humanity, or posthumanity. Such visions occur regularly in science fiction and theological writings, but appear rarely in philosophical texts. On the assumption that it will be science; that is, what we currently regard as natural science, that offers us serious possibilities of altered human nature, it will be the task of ethics, broadly construed as involving social and political philosophy, decision theory, economics, and related thought, to weigh these possibilities.

There are two immediate concerns regarding such a proposed task. First, it will likely be countered that humanity has been evolving since the dawn of civilization, so that whatever the major tasks of ethics they either have remained unchanged or, if they have changed with changing socio-historical contexts, there is no further radical departure to contemplate. If there is a correct ethical theory it is trans-historical, and if there is no timelessly correct ethical theory, then the truths of ethics, insofar as ethics can convey truths, are contextual.

Second, suppose—contrary to the first concern—that ethics will need to confront questions it has not yet considered, such questions arising from proposed radical alterations in human nature. Suppose further that contextual ethics is rejected as being relativistic. Then by what criteria can such questions be decided? Some might have ready responses to this question. However, assume the correctness of my above-stated claims that ethics has not yet delivered a successful theory, nor can it deliver such a theory as human nature is currently constituted. If, on the promise of future science and technology, a possibly evolved humanity expects as much from the study of living well as it currently expects from the study of natural phenomena, then ready responses from our current standpoint; that is, what we currently know and are, are suspect.

Much of humanity in the so-called "developed sector" expects continuing insights into the nature of things from natural scientists. Expectations regarding the deliverances of social and behavioral scientists are markedly less robust. For humanity to honor ethical theory to the extent that it epistemologically honors natural science, the situation of the social and behavioral sciences will likely need to change. One possibility is that the social and behavioral sciences will in some fashion reduce to the more successful natural sciences. Another possibility is that the social and behavioral sciences will become more respected as a result of future transformative discoveries. There are of course intermediate positions here, involving some manner of overlapping of the various social and behavioral sciences with natural sciences. Such overlapping has been a fact of neurophysiologically oriented psychological science, to the extent that neurophysiological phenomena are correlated with behavioral phenomena.

Many of the most influential statements of ethics—for example, Plato's, Aristotle's, Spinoza's, Kant's, and Mill's—have an applicability that apparently transcends their historical situatedness. To the extent that this is so, any emotivist, existentialist, deconstructionist, or similarly truth-denying metaethical statements are suspect. Contextualism is similarly suspect. There is the hard fact that the former classical theories are often prima facie inconsistent with one another. But perhaps they are approachable with a conciliatory attitude. Kant, for example, might be read as emphasizing respect for the power of thought that resides in rational creatures, whereas Mill might be read as focused on the joy that attends the higher pleasure of thinking.

Yet the contextualists have their point. The specifics of various situations at least sometimes elude the generalities of the great ethical theories. The truth-deniers, such as the non-cognitivists, can make strong cases out of various examples of seemingly irreconcilable, conflicting demands. So we are left with the unhappy fact that ethical theory is not a settled matter.

If, as I have been urging, this unhappy situation is partially due to our incomplete knowledge of human nature, and partially due to irreconcilable conflicts in that nature, which conflicts are mirrored in conflicting theories (for example, classical-liberal individual acquisitiveness versus Marxian social generosity), then ethics to this point is at the least incomplete. Whether it is successfully completable will in part depend on our ability to alter human nature. There are two "alteration schemes" worthy of consideration. In one, the desirable and undesirable traits in people are identified, and future generations are fitted appropriately. Quickness to anger, for example, seems a trait that has outlived its evolutionary usefulness. Assuming we develop the means to expunge such traits, we need to consider carefully the social purposes such traits serve, distributed as they are throughout the human population. Such consideration will likely involve concepts and theories that have not yet been formulated.

Whether such personality-engineering will produce the near-ideal state of affairs referred to above is a question that I do not believe can be answered satisfactorily at this time. Kant suggested a manner in which we ought to behave towards one another, yet he was aware of the conflict in our natures between the call of duty and the call of desire. Kept within the confines of human desires, as humans are presently constituted, and kept within the confines of a limited life-span, such personality-engineering might be inherently too limited to achieve what I am positing as the ultimate goal of an ethical theory—to delineate and elaborate the conditions of near-ideal human existence.

On the supposition that the above indicated personality engineering is insufficient to realize this ultimate goal, the question arises as to whether the second "alteration scheme;" that is, more radical alterations of human nature, could occasion the goal. Consider once again the suggestion that humanity evolve into a species of sexless, childless, cognitively-motivated beings of enhanced intellectual abilities with indefinitely extended life-spans (which apparently necessitates childlessness). Suppose further that these evolved beings have desires and satisfactions that are unlikely to conflict, and to the extent that interpersonal cooperation is needed, it is typically granted ungrudgingly.

I suspect that many, if not most, current adult humans would not favor this vision over their present lives. Allowing current choice as determinative of value, the above outline is not that of a near-ideal state of affairs for humans. Yet there are those, myself at least, who would argue that such a scheme would allow a near-ideal life for its participants. In support of my preference, recall Plato and Aristotle's various ruminations on the advantages of the contemplative life; that it can be pursued without surfeit, that it does not necessitate conflict with others, that it demands scant resources (at least for the immediate act of thinking).

To such reasons I would add that knowledge and wisdom are held in high esteem by various cultures, albeit with significant differences among them as to the proper domain of knowable things. It is possible, if not probable, that as our understanding of what we are progresses along certain lines, we may as a species collectively decide to migrate physiologically to a terrain that exemplars among us have afforded us glimpses, less by their expressed thoughts on the matter than by the quality of their lives (for example, Einstein) as these have been publicly viewed. If there comes to be a worldwide acceptance that dispositions such as Einstein's are preferable to those of most humans, and if it becomes feasible to endow humans en masse with such dispositions, then it is possible that humans en masse will choose such endowment.

It might be said that a world of Einstein-like temperaments is not possible, in that in such a world conflicts negating those temperaments would arise. Accepting this reply, the above sketch of a worldwide conversion of humanity to such temperaments is intended as a prod to intuitions. It seems likely that more far-reaching alterations are necessary for the general avoidance of conflict and concomitant flourishing of the life of the individual. What these alterations are I have sketchily indicated above. Of course other choices are possible. It might be urged that humanity's favored future is in a loss of individuality in favor of a collectivity of the sort variously portrayed in science fiction venues. It might even be urged, and doubtless it has been, that humanity's proper choice is a controlled "dumbing down," thereby allowing a more "natural" existence.

If there is no consensus regarding the future direction of humanity, yet the possibility of transformations of human nature is increased through scientific and technological advances, then the question "whither humanity" will be at the forefront of ethical concern. Among the possible responses to this question are the following:

1. The discipline of ethics cannot answer this question. Ethics assumes the human. The posthuman is outside the domain of ethics.

   This is analogous to the response occasionally given to a question raised in the domain of thermodynamics regarding the second law (sometimes termed "the law of increasing entropy"). The question concerns the situation of the entire universe with regard to that law; is the universe as a whole increasingly entropic? The reply sometimes given is that the second law applies to thermodynamically isolated systems, and given the assumption that the universe is (perhaps by definition) not in isolation from anything else, the second law is inapplicable to the universe as a whole.

   While not responding to the correctness of this position either as a matter of ethics or a matter of thermodynamics, there is no denying the intuition some have that such a response fails to satisfy the curiosity prompting the question.
2. Whatever emerges in time is justified as what humanity has chosen.

   This ostensibly nonethical response is compatible with some ethical theories, perhaps most easily those of a libertarian flavor. The libertarian assumptions are that the human condition over the relevant time period meets a threshold of informed and unfettered choice-making circumstances, and that the proper (best? most justified?) choice in such conditions is the choice actually made.

   Considered under this libertarian justification, the above response is subject to the intuition that some outcomes, however "freely" chosen, may prove dystopian from both current and post-choice standpoints.
3. There are now, or there will emerge, pointers to the best choice.

   The current lack of what I have termed a "near-ideal state of human affairs" does not entail lack of indications of posthuman utopia, but I am doubtful that any set of such indications are entertained in a manner resembling that in which entrenched natural scientific beliefs have worldwide acceptance. My doubts here do not preclude speculation based on present circumstances, and such speculation will be pursued in succeeding chapters.

> As to pointers at some future time, it is my hope that these will emerge as scientific research discloses a more detailed picture of the psychological aspects of human nature, and presents humanity with biophysical and psychological possibilities that are not currently envisioned.

Having the ability to transform its members into something other than human, there is the choice of humanity to adopt some such alternative, or to remain human. I have indicated above that I find strong reason for humanity to evolve to posthumanity, but as I stated above I do not believe that, given present information, this choice can be made with epistemic confidence. We may choose to remain human, in some manner either progressing, regressing, or maintaining a stasis. Whichever of the latter three alternatives are chosen, there are historical instances of significant human populations exemplifying that choice.

The choice to remain human, or evolve to posthuman, may be closer than one might suppose, given present and foreseeable scientific discoveries. The possibility of such discoveries accentuates the problem of fixing criteria to guide the choice. There are various philosophical discussions of apparently similar choices, such as the choice of what sort of life one should lead. Both Plato and Mill struggled with questions regarding the life of pleasure: Plato's *Republic* Book IX examines the question of being whether pleasure in itself is an appropriate life-goal; in Mill's *Utilitarianism*, whether the "higher" pleasures are to be preferred to the "lower" ones. One of Plato's several responses matches Mill's singular response: those who have experienced the range of alternatives are the appropriate arbiters.

However, there is a significant difference between choosing human alternatives from a human perspective and choosing whether to continue a human perspective or not. The latter choice is one ostensibly disallowing arguments such as the one above; namely, from acquaintance with the options in question. The prohibition may not be absolute, however. It may be that aspects of the posthuman already exist in various humans, and this may provide the above argument with a foothold. Suppose, for example, persons of astounding intellectual talent. If their talent is not of the restricted, idiot-savant sort, such persons might provide the rest of humanity with some idea of what it would be like for all to have such talent. Whether such persons are already among us is partly a matter of refining what is intended by "astounding intellectual talent" and partly a matter of testing for such talent.

Alternatively, whether humanity has intellectual, moral, aesthetic, or other such *übermenschen*, it might be impossible to generalize imaginatively their number in a convincing manner. Still, there are haunting questions of the form, "what would a society of Einstein-endowed, Einstein- or

Ghandi-temperamented individuals be like?" To the extent that we are drawn to such possibilities, some manner of an experience-based choice might be operating. Perhaps we are comparing a "standard picture" of ourselves with a picture of ourselves at our "best moments," and identifying the latter with our images of individuals such as Einstein or Ghandi.

Whether the question of choice is as to which form of evolved, posthuman life, or whether to evolve rather than incrementally improve our human form of life, the question is afflicted with the uncertainties of decision-making based on inadequate information. I have suggested that what humanity is seeking is a near-ideal set of living conditions, and implied that we will know this state when we arrive at it, or perhaps when we are approaching it. If this broad statement is generally correct, then it seems a major task of ethics will be to clarify our aims as we approach this longed-for state. The philosophical foci, be they anything resembling current analytic or continental practice, might continue as now or, perhaps at the other extreme, the impact of attaining this near-ideal state will banish ethics from human or posthuman practice in the manner in which philosophy has largely been banished from the natural sciences, in which case, ethics has a limited future. Alternatively, if that self-satisfied state that I have speculated for humanity is not an end-state so much as a baseline condition for all that follows (analogous to what Marx refers to as the end of the prehistory of humankind[7]), ethics might remain as a—possibly the—deliberative discipline regarding the choices that emerge.

In offering the opinion that humanity, as it progresses, will and ought to consider evolving to an other-than-human condition, I face the challenge that it is not the primary business of ethics, if it is the business of ethics at all, to speculate about future possibilities. Taking this challenge literally, I reply, "So much for Plato's *Republic* or Nietzsche's various preachments, among other respected authors." On the other hand, suppose what is meant in this challenge is to the extent that ethics addresses the human condition, it is the current and foreseeable human condition that is of concern. Then more needs to be said. I conclude with the following minimal remarks.

Peace on earth has not been realized, nor have various needless occasions of human suffering unrelated to warlike conditions been eliminated. There is much to be done in the present. If the theory and practice of philosophical ethics has not led to the fullest realization of human happiness under present circumstances, this cannot be attributed largely to such theory and practice. As I have noted, our current scientific ignorance of the deep facts of human nature, together with the more obvious deficiencies and conflicting tendencies of such nature, serve to prevent a near-term significantly better world. Against these impediments ethics does what it can.

If it is being said, however, that ethics is and must be a largely academic conversation, that it should not be burdened with the historically situated, socio-political problems that confront humanity at large, then I submit such a statement shows its worth on its face. That evaluation being the case, I conclude that humanity's future will determine a portion of ethics' future, and the latter's future will not completely resemble its past.

NOTES

1. There are a number of recent works of immediate relevance to this topic. Although not written by professional philosophers, they have philosophical content. Included among such recent works are: Joel Garreau, *Radical Evolution* (New York: Doubleday, 2005); Ramez Naam, *More Than Human* (New York: Broadway Books, 2005); Keith Stanovich, *Robot's Rebellion* (Chicago: University of Chicago Press, 2004); Simon Young, *Designer Evolution* (New York: Prometheus Books, 2006).
2. Thomas Kuhn, *The Structure of Scientific Revolutions*, 3rd ed. (Chicago: University of Chicago Press, 1996).
3. Jean-Paul Sartre, *Existentialism and Human Emotions* (New York: Philosophical Library, 1957).
4. Examples of such include: Peter Van Inwagen, *An Essay on Free Will* (New York: Oxford University Press, 1983); Daniel Dennett, *Elbow Room* (Cambridge, Mass.: MIT Press, 1984); John Fischer, *The Metaphysics of Free Will* (Cambridge, Mass.: Blackwell,1995); Derk Pereboom, *Living Without Free Will* (New York: Cambridge University Press, 2001).
5. The most well-known defense of the social usefulness of such reactivity is Peter Strawson's "Freedom and Resentment," *Proceedings and Addresses of the British Academy* 48 (1962): 1–25.
6. Sigmund Freud, *Civilization and Its Discontents* (New York: J. Cape and H. Smith, 1930), chapter V.
7. Karl Marx, *A Contribution to the Critique of Political Economy* (New York: International Publishers, 1970), 22.

Chapter 2

Minds and Related Matters

What are those successors of current humans who will live indefinite lives, assuming that there will be such? The question is multi-dimensional in that there are considerations of embodiment and considerations of mentation. Further, whatever such successors are at some time, they may later evolve into something else.

What we might become is in part dependent on what we are now. We know much about ourselves. That is, we know much about our species-wide characteristics. But, as was stressed in the preceding chapter, regarding species-wide characteristics there is much we do not know about ourselves. There is nothing that can be done here to allay that lack. Moreover, it is not only a matter of filling in such details as the workings of the brain, as difficult as those details might prove to be. Consensus is lacking on various metaphysical contours of what we currently are, and have been for some time.

It is with the above in mind that I offer a sketch of selfhood here. If what is said above is correct, this sketch will necessarily suffer some of the same limitations as any past or current account of the self, whether given at length or, as here, indicated rather than elaborated. In particular, I acknowledge three features of what follows: the account may be fundamentally mistaken; to the extent that conclusions are derived, they are more soft than certain; little of the account, if any, is original. It will be seen that, as with any competing story, this account has consequences regarding the overall discussion of transhumanism.

To begin, of the various possibilities regarding the relation of mind and body, I am an epiphenomenalist. By "epiphenomenalism" I intend that position regarding the mind-body problem which holds that at least some of what is termed "mental" or "psychological" is nonphysical, and further that this nonphysical portion of the mental has physical causes and itself has no causal powers. If the reader is unschooled in this position, the following analogy, although somewhat misleading, may be helpful. The system of a movie projector running a film and projecting images upon a screen is the cause of those images. Assume the images depict some activity, say, a naval battle. Cannons are firing, ships are set aflame and sunk, sailors are screaming. The film depicts causal activity galore. Yet viewed in one manner, none of what transpires on the screen is causal. The images of cannon fire are not causing the images of explosions, the images of flaming ships are not causing the images of the sailors' frantic activity, and so forth.

The analogy is hopefully transparent. The caused-but-lacking-causal-powers screen images correspond to the mental epiphenomena, whereas the system of projector and film correspond to the physical causes of the epiphenomena. The imperfections of the analogy include: both the film and projector system and the projected images are physical things, and there is an amount of causality in the projected images inasmuch as they causally impinge on the screen; images are visible things, while much that I consider epiphenomenal is non-visual.

It is tempting to identify that which is nonphysically mental with that which is conscious, so that psychological epiphenomena are conscious phenomena and conversely. I am hesitant to acknowledge this identification due to vagaries and ambiguities of the term "conscious." "Consciousness" has affinities with terms such as "awareness" and "sensitivity," and the latter two terms, especially the last term, are often used in contexts in which they designate physical matters. This caveat presented, I am willing to consider "consciousness" as referencing what I consider the various nonphysical aspects of the mental.

But what is the reason for considering any aspect of the mental as nonphysical? There are at least several general responses to this question. The response I find convincing is that those aspects of the mental that are nonphysical are not to be found in physical space. Indeed, I take the absence from physical space of certain phenomena, typically—but perhaps not exclusively—a subset of phenomena deemed mental, to be criterial for the nonphysical nature of such phenomena.

Those familiar with the overall mind-body discussion know that the relevant literature easily fills a substantial bookcase, if not more.[1] I do not intend my discussion to be decisive. Knowledgeable readers are aware, however, that those who do so intend their "solutions" to the mind-body problem are countered by others with similar intentions and opposed "solutions." All this taken for granted, it yet behooves me to offer a modest defense of my "solution" to this outstanding problem. I will offer reasons for my position and attempt to fend off objections. But I am more interested in the implications of my position with respect to the major focus of the following chapter, which is the passage of the human to the posthuman—that is, transhumanism.

The notion of physical space, upon which my position on the mind-body problem relies heavily, is not a simple notion, unencumbered by natural-scientific considerations. Relativistic physics has it that the geometry of physical space is influenced by gravitation. There are aspects of quantum theory that allow nonintuitive conceptions of extremely small volumes of physical space. All this granted, for the purpose of this discussion I propose a simplified notion of physical space. Physical space, simplified, is the medium in which all physical phenomena is contained. Or rather, together with physical time, physical space constitutes that medium.

I do not propose that the epiphenomena with which I am concerned do not occur in physical time. Of course, it is only the correlated brain activity—to the extent there is correlation—that is subject to such direct temporal dating as can be given distinct brain activity. I am supposing that the epiphenomena caused by such brain activity can be approximately dated with respect to the dated brain activity.[2] If there are problems with this idea of indirect dating, aside from the precision of such indirect dating (assuming the imprecision is confined to intuitively tolerable range), I retract the supposition that such epiphenomena occur in physical time, although there must then exist an "internal" or "subjective" time in which these occur. I find it unnecessarily complicating to introduce these notions into the discussion, unless evidence and reasoning with which I am currently unfamiliar necessitates that I so proceed.

If one endorses epiphenomenalism, why not interactionism? That is, why not suppose that nonphysical mental phenomena have causal powers such that these phenomena interact with the physical realm, particularly with the organism said to experience such nonphysical phenomena? I concede that such is possible. I further believe that such interaction is a necessary, but not sufficient, condition of the realization of such notions as autonomous agency, and genuine freewill. I shall not defend these further beliefs here. Given that I am not an interactionist, and that I believe interactionism is a necessary condition of the realization of the above notions, it follows (assuming my rationality) that I do not believe human beings are autonomous agents, possessing genuine freewill. Since I believe that such freewill is a necessary condition of moral responsibility, it follows that I do not believe human beings are morally responsible for their actions.

If I am correct in these preceding beliefs, there is much at stake in the affirmation or denial of interactionism. My denial is based upon a standard reason; interactionism is in violation of the conservation of momentum. Put more generally, if physical events in the brain are fully determined by prior physical events, then any alteration of such determination by nonphysical phenomena constitutes a violation of physically lawful causality. There is significantly more to be said. In particular, the antecedent of the more general formulation above is not without challenge. I repeat that I do not place full confidence in my denial of mind-body dualistic interactionism. Much of what I will maintain is compatible with the metaphysical stance of interactionism, especially if it is granted that the existence of genuine freewill demands more than plain interactionism. Such interactionism, after all, could fall under invariable laws that link the nonphysical to the physical. If this linkage is not so, further reasons are needed to support its denial.

I place more confidence in my argument for psychophysical dualism. If what is physical must exist in physical space and at least some phenomena typically considered mental do not exist in physical space, it follows that some mental phenomena are not physical. The likelihood of these phenomena being caused by brain activity rules out positions such as mind-body parallelism. The remaining possibilities are epiphenomenalism and interactionism, and of these two possibilities the problems of the nonphysical mental causally interacting with the physical incline me toward epiphenomenalism.

The argument given above appears open to the charge of question-begging. If mental phenomena are brain-located (or, more broadly, brain-, central nervous system-, and bodily behavior-located) then they are in physical space, and hence are physical.

Are there reasons against this counter? Consider the example of dreams. If we really dream in images, and if these images—their conceded physical causes aside—are not to be found in physical space, then psychophysical dualism is a matter of fact. Can it be argued either that we do not really dream in images or that these dream images are not in physical space? If our dreams are populated with images and these images are nowhere to be found in the physical space inside our skulls, then it is difficult to maintain that these images are in the same space occupied by our physical aspects. The better physicalist strategy is to challenge the claim that these images exist as images, rather than as items that can be maintained to exist in physical space.

Such a strategy initially seems impossible. Many of us, if not all, have had vivid dreams on occasion. As Descartes puts it:

> How often has it happened to me that in the night I dreamt that I found myself in this particular place, that I was dressed and seated near the fire . . .[3]

The physicalist's situation here is not hopeless. If one is inclined to a sort of realism that disallows the existence of vague or ambiguous circumstances in reality, maintaining instead that any ostensibly vague or ambiguous circumstances are due to epistemological—rather than ontological—conditions, one must face the physicalist's insistence that dream images, as well as any claimed items of nonphysical consciousness, are significantly vague. For example, while one has little difficulty counting the spokes of one's present and in-view bicycle wheel, counting the spokes of one's supposed memory image of a childhood bicycle wheel is a different matter.

The upshot of the physicalist's denial of the existence of such items of dualistic consciousness in a nonphysical realm (or realms) of space (or sounds, tastes, odors, and so forth) is the physicalist's explanation of these phenomena. The typical physicalist account references brain activity of a sort casually termed by the layperson "brain traces." That is, the brain activity that happens during actual perception of physical entities is in some similar manner occurring in the absence of such actual perception. This "shadow" brain activity is suggesting the perceived presence of actual physical entities without there being anything such as images present anywhere—including in supposed nonphysical consciousness.

The claim that one's supposed nonphysical conscious items are vague beyond acceptance as real entities relies on various assumptions. Perhaps most immediately, it relies on a form of realism that itself relies on the principle of bivalence; that any claim made about a state of affairs is either true or false. One's remembered bicycle wheel, for example, has some number of spokes. Yet no number can be confidently given. While this criticism is possibly negotiable by allowing that any claimed specific number of spokes in the considered situation is false, such a tactic runs counter to insistence that one's supposed image is minimally constituted of a wheel with discrete spokes.

There are two other assumptions the physicalist must hold in this context. One, that the reports of the introspecting subject are honest. In the present state of those sciences relevant to this discussion, the subject can claim a detailed "view" of conscious images even if this claim is bogus. If in the example of the remembered bicycle wheel the subject claims the imaged wheel has twenty-eight spokes, which the subject clearly "perceives," it will not do to remind the subject that the former wheel really had thirty-two spokes. It is open for the subject to reply the image in consciousness clearly has twenty-eight spokes, regardless of the actual number of spokes of the former article. For the purpose of this discussion, let introspective honesty be presumed.

The second assumption the physicalist must hold, given the flow of this discussion, is that no one is genuinely capable of having images (or sounds, odors, and so forth) in their consciousness that are as clear and detailed (and, if necessary, coherent-in-time) as their actual perceptions. This is of course an empirical matter. I propose not to found any claim on the possibility, genuine though it might be, that this assertion of the physicalist is false. The reason for my proposed generosity is contained in my appeal to what I consider obvious: we have dream images, we have imagined images, and we have recollected images. These images, vague though they are, are at times apparent to us, and they are not in the space of physical things.

Those of realistic attitude who accept the existence of items such as dream images can take refuge in Aristotle's dictum of requiring as much certainty as the subject allows. Whether genuine realism permits any vagueness among items said to constitute reality is a topic of much discussion.[4] Although I consider myself a realist in these matters, I do not believe the complete precision of entities is maintainable, except perhaps in an ideal reconstruction of some sort. There are problems related to Zeno's paradoxes. Less recondite are the problems of specificity attending the location of such entities as gas clouds. And in the less than mathematically continuous space of dreams, to say nothing of whatever medium(s) imagined sounds and odors reside, numeric precision is unrealizable. I must content myself with the following: I would sooner surrender the label of realist than surrender my belief in the existence of dream images, imagined sounds, and so forth. Such a declaration does not amount to decisive argument, but I see nowhere further to go.

The physicalist has other possibilities. There is the mysterious nature of the causal nexus from the physical to the nonphysical. To my knowledge, the only reply to questions such as, "where is the point of action from neural activity to nonphysical images?" is simply that such causality happens. That question is answered in similar manner to the grand ontological question as to why there is something rather than nothing. There simply is. Provided there is no expenditure of energy in such causality, the epiphenomenalist must be content with knowing that there is no violation of fundamental physical principles.

The epiphenomenalist must be carefully consistent here. Nothing in nonphysical consciousness can be allowed to influence underlying physical states of affairs. If, for example, the subject is extremely snake-phobic and happens to imagine a snake for no apparent reason, and if the subject then becomes visibly agitated, the epiphenomenalist cannot suppose that the nonphysical image of the snake is in any manner responsible for the subject's agitation. If a subject reports, upon query, as to the content of the subject's nonphysical consciousness, the epiphenomenalist cannot suppose that it is the actual content of that consciousness that is in any manner causally responsible for the subject's public report. This sort of consideration lends epiphenomenalism an air of incredibility similar to that attending mind-body parallelism. But consider an interactive computer game as it unfolds on a monitor. The images onscreen are not invested with causality vis-à-vis the unfolding of the game-events, although one will often speak as though they are. While these images are having some causal impact on the screen they inhabit, in terms of the actual underlying physical causes unfolding in the computational machinery—including the display producing machinery—they are epiphenomena. Appearances to the contrary, the monster does not explode because shot with the ray pistol in the hands of the screen soldier. In the physical sense of "explode" the monster doesn't explode at all. I trust the analogy is patent.

The matter between the epiphenomenalist and the physicalist is not settled. It may come to pass that physical theory includes remarkable concepts of space which allow a physical response to the question of location of dream images (and even imagined sounds, odors, etc.). There are current intuition-defying quantum-theoretic hypotheses regarding microspace. The reader is urged to consider such a future possibility in light of what follows. If what I will urge regarding the value of psychophysically dualistic consciousness be viable under a fully physicalist assumption, then much of what I have said above is at best superfluous, given my aim.

There is an important philosophical area where I am in general agreement with the physicalist. I do not suppose that an appropriate model of perception and the external world need be dualistic or idealistic. That is, of the three general models of perception and the external world—phenomenalist, dualist, direct realist—I favor the last of these. While I cannot adequately defend my choice here, the relevance of this choice to the above discussion merits notice.

The physicalist may seek to maneuver the psychophysical dualist into a difficult position in the following manner. Suppose the epiphenomenalist maintains that one's perceptions are items of one's nonphysical consciousness. Then the consistent epiphenomenalist subscribes to either a phenomenalist model of perception and the external world, or to a (representational?) dualistic model. These days, phenomenalist models of perception are tainted by association with logical positivism, and dualistic models have classical difficulties countering charges of scepticism and solipsism. As quickly presented here, such challenges are hardly decisive. One can suppose there may be adequate defenses of the choice of either a dualist or phenomenalist model of perception. Prima facie, however, a direct realist model of perception avoids having to defend against these challenges.

This avoidance comes at an apparent price. In joining an epiphenomenalist position on the mind-body problem with a direct realist take on the problem of perception and the external world, I am maintaining that, although there is nonphysical consciousness, perceived items are not part of that consciousness. Indeed, they are physical things. In the case of visual perception, they are the physical things we see. But, surely, consciousness has something to do with our perception! Correct, and correct in two ways. It needs be remembered that the term "consciousness" is often used in an ontologically neutral manner. This is especially the case regarding usages in which "conscious" is roughly synonymous with "awake" or "aware," and in contrast to "unconscious." Further, nonphysical consciousness is often attendant perceptual activity. Perceptions may occasion emotions, for example, and the "felt" qualities of these emotions are arguably items of nonphysical consciousness.

To press difficulties: often one is unaware of objects centered in one's visual field, yet it can be assumed that, barring exceptional circumstances, a physical image of the object is present on one's retina. One way of describing this situation is that the image is present on our retina, but absent from our consciousness. Hence, what we do see is present in consciousness. If this manner of description holds, however, then the epiphenomenalist apparently must choose between phenomenalism and representational dualism. In response, another way of describing the situation is that the object; that is, the physical object, simply is not seen. If one desires more than wordplay in the direct realist's response, consider the following. Seated at the end of a long, rectangular table, one sees that table as trapezoidal. That is, if one attends to the "look" of the table—which one likely knows to be rectangular—it "looks" trapezoidal. The image on the focusing plane of a camera will be trapezoidal. As J. L. Austin would say, "Things that are rectangular need not look rectangular."[5] We can go further. A tape measure of satisfactory length will reveal that all opposed sides of the table are equal, as well as the two diagonals. The act of measuring the table involves looks (typically, correlations between marks on the measure and table edges). We are thus presented with competing "looks." In declaring the table rectangular, we need not be understood as speaking of an unseen *ding-an-sich*. We are not only talking about the physical table we have seen, we are talking about our seeing the table as rectangular. While there are competing "looks" there are not competing ontological perspectives. Our choice of predicates for the physical table is based on valuing one manner of looking over another. The other manner of looking has its place. Painters, for example, value it. But we often prefer a more perspective-invariant description of how things are.

The reader is likely impatient. Little or nothing that has been said above is directed toward the topic of transhumanism. Yet what has been defended, psychophysical dualism, is pertinent to the discussion of transhumanism, as I will show shortly. Regarding perceptual realism, the importance of that notion to the discussion will be argued somewhat later in this chapter.

I maintain that, on the assumption of physicalism, many phenomena that are gathered within the category of consciousness no longer exist. If we are solely physical beings, we lack what are typically termed "feelings." We lack the rich, inner world of imagined sights, sounds, odors, tastes. We also lack the experiences generally referred to as "talking to ourselves."

Physicalists of course deny this. According to them, all such phenomena are accountable as physical phenomena. Various reasons given by physicalists for such encompassing have been canvassed above. An additional consideration, not mentioned above, is the theoretical identification of seemingly disparate items, such as lightening flashes with a discharge of electrically charged particles, or of gas concentrations with molecules having a significant amount of freedom of movement. In part, such putative identifications are actually reductions from apparent macrophenomena to the supposed exclusively real underlying microphenomena. The reductive component of this physicalist maneuver will not be discussed here, although I do not accept the honorific use of the term "real" as applied to the micro-parts at the expense of the macro-whole. In any case, this sort of physicalist reply does not answer the epiphenomenalist's argument that some mental phenomena do not reside in physical space. The space, and in rough outline the places, in which lightening flashes and gas concentrations occur is the same space as that of their theoretically identified microcomponents.

I believe it is an open question at this time whether any embodied entity could possess the cognitive abilities of homo sapiens sapiens, or of any mammal, including being-in-the-world-know-how, and lack epiphenomenal consciousness. I have observed demonstrations of robot hands and robot heads having eerily human behavior. In the latter instance, I have seen a humanoid robot head having expressive movements which are activated in reaction to what the robot "observes." The robot has vision and auditory systems with pattern recognition software enabling it to identify, with passable accuracy, various emotional manifestations in humans. When confronted face-on the robot responds with complementary emotional expressions, such as sadness or anger. The effect can be startling. Yet we are reasonably certain that the robot head lacks genuine epiphenomenal consciousness. What seems less certain to various physicalists is whether a robot with broad-spectrum cognitive abilities and motor skills roughly equivalent to those of a human could be said to lack all that is indicated by the term "consciousness." The epiphenomenalist's concern will be whether such a being has a nonphysical "interior" life. The physicalist will reply that the assumed capabilities are all that is the case regarding a mental life, and that all the particulars of such a life occur in the same arena as do all physical phenomena.

Perhaps I am mistaken in my psychophysical dualism. I am less concerned with being correct on this point than on the following. Whatever humans transition to, those transitional entities ought to have something of the sort I am terming an "interior life." If they possess no more "felt" feeling than robot heads of the sort mentioned above, then regardless of their other capabilities, there is little to recommend them as a future for us. Given my epiphenomenalism, I am strongly recommending an epiphenomenalist future for us. At the least, I am strongly recommending a future in which we enjoy our accomplishments. And I do not identify enjoyment with enjoyment-behavior. I find it mistaken fully to identify enjoyment with anything physical, although I believe enjoyment has physical causes and physical components (that is, there is a physical side to enjoyment). I further believe that terms like "enjoyment" are occasionally useful, albeit crude. There are those who in the course of their disciplines specialize in vocabularies of consciousness, and until such time—if ever—that better (more precise?) treatments of consciousness are available, I defer to them.

The preceding discussion posits consciousness as a necessary condition of worthwhile transitioning. There are some additional posits I want to make. Generally, these are posits of realism. The sense of "realism" I intend is, broadly, that there is a world of typified objects, events, states of affairs, and that this world, together with its characteristics, is independent of our epistemic relations to it. There is an item of faith that is implicit in this sort of realism; namely, that appropriately endowed cognitive agents can come to some form of understanding of this world.

The vagueness and otherwise incompleteness of these notions, as just presented, is obvious. Some small amount of detail will be supplied by the ensuing discussion.

There are proposals that humans evolve into entities existing in virtual reality. Let us put aside for the moment the extraordinary difficulty in creating and presenting a detailed and coherent virtual reality acceptable to minds possessing human cognitive powers and human experiences. Variations of these proposals include remaining in bodily form while uploading our active minds into such a reality, versus losing our embodiment except insofar as we exist within physical computational machinery as virtual entities in virtual realities. Such visions admit of further detailing. For example, we might maintain our physical selves and enter virtual reality on occasion.

Regarding any such scheme, I oppose our sojourning in virtual reality in any manner other than possibly a sometime-holiday or vacation. My reason for such opposition is that I am a realist in the above-given sense. I also hold that, at our highest, in our present human circumstances, we are creative and cognitive beings. This last proposition is in need of elaboration and defense. Further, assuming an adequate elaboration and defense of the preceding assertion, the proposition that we should evolve into posthuman beings whose hallmarks are their cognitive and creative abilities is in need of elaboration and defense.

Assume for the moment that these propositions are adequately elaborated and defended. Allow one further main assumption—that the above-referenced cognitive and creative abilities do not conflict with each other. Allow one sub-assumption—that the notion of conflict in the preceding sentence can be adequately specified. These assumptions granted, our posthuman future is understood not to be best spent in virtual reality.

Defending realism requires another book, and since many such works exist on all sides, the project is of dubious worth. If one adopts the Socratic view that the only successful defense of a thesis is one in which opponents are committed to propositions which imply the thesis, I harbor hope for the triumph of philosophical realism, but there is the possibility that I will find that my own commitments entail a contrary conclusion. Relaxing the requirement of realism, as loosely sketched above, what is of concern here is the acknowledgment of a world external to consciousness, towards which conscious beings of certain sorts might have cognitive relations. If these cognitive relations are minimally of the sort that exist between the practitioners of modern natural sciences and their domains of concern, I am satisfied. There are well-known philosophical controversies regarding the proper account of such cognitive relations. Still, there are few practitioners of modern natural science who take their areas of concern to be virtual realities (with the possible exception of some quantum theoreticians).

While I may have succeeded in indicating my philosophical discomfort with the suggestion that we evolve into beings who largely inhabit virtual realities, I have likely occasioned some ill-will by what I have said above. I have apparently privileged natural scientific cognitive activity, and not everyone bears this attitude. Some believe that there are other modes of cognitive thought of equal or higher value than that of natural science. And the assumptions made a few paragraphs above are minimally in need of a modicum of elaboration and defense.

Of course, we cannot engage in cognitive activity if we do not exist. Existence being a necessary condition of engaging in any activity taking place in physical space and time, an acknowledgment of the material conditions of our existence is appropriate. Such an acknowledgment should reach beyond minimal food, clothing, and shelter requisites. The outlines of the forms of living that constitute our current possibilities are significantly wider than those of our Paleolithic, genetically equivalent ancestors. Admittedly, in an intuitive sense of "depth," the sensitivity of these ancestors to their immediate environment was likely deeper than ours. Similarly, in moving from, say, nineteenth-century middle-class life (for those in that timeframe fortunate enough to be so stationed) to current middle-class life, there has been a complexity of both loss and gain. I believe the gain on the whole outweighs the loss, but I shall not argue the point. If one does not accept this, one will question some of what follows. All sides will hopefully accept the proposition that natural-scientific thought, with attendant technological developments, has been a major factor in the transformation of life in the industrialized countries from the nineteenth century to the present time.

For the reason given in the preceding paragraph, I emphasize those forms of cognitive activity that constitute what I broadly term "natural-scientific thought and practice." I do not mean to suggest that such are the only forms of cognitive activity, or that other forms of cognitive activity have little or no importance. I do mean to emphasize the creative components of natural-scientific thought and practice. Once one goes beyond "formula plugging" or otherwise routine applications, science and technology are creative endeavors. Momentarily adopting Kuhn's bifurcation of scientific practice into "normal" science and "revolutionary" (or "paradigm-introducing") science, the former form of practice—as well as the latter—often demands creative thought, whether by the researcher, the practitioner, or the academician. Scientists and technologists are often focused on goals to the extent that they fail to note the ingenious, or at any rate novel, steps that they have taken en route. Such novelty includes reformulation and alternative sequencing of preexisting routines.

But creativity is not solely the province of the scientist and technologist. Anyone having a purpose who deviates from a routine is being "creative" in my sense of the term. One engaged in the mundane activity of housecleaning, who considers and executes a more efficient manner of removing accumulated grime under the refrigerator, is being creative. Of course, if profound metaphysical determinism is either fully or largely the case, creative activity does not transcend the ontological routine of natural law. In such case, ostensibly creative activity must be so considered from a more limited perspective.

As employed here, the notion of creativity is most commonly associated with aesthetic endeavors. Insofar as artistry is other than routinized craftwork it merits the title of creative activity, as the concept has been loosely specified above. We might question, however, whether the aesthetic life, or even aesthetic moments, are appropriate possibilities for posthumans. Such questioning seems pointless if directed at a technological-scientific mode of life for the reason given above; namely, that the material basis of our present lives—and likely of the lives of our descendants—is secured through appropriate scientific and technological practice, together with attendant techniques of production and distribution.

These matters are not as neatly decided as has just been presented. The "attendant techniques" spoken of in the preceding paragraph, especially those of distribution, involve normative deliberation. If such deliberation becomes scientific in the manner suggested in the previous chapter, then citing the normative nature of productive-distributive choices poses no additional challenge to the value placed on a science-centered posthuman existence. If such deliberation cannot be made a matter of science, however, additional complexity ensues. Resolutions are possible. If, for example, there results a general, worldwide consensus regarding the value of various sorts of life, then however nonscientifically this consensus has emerged, it remains for science and technology to secure the material basis for such life. There may of course be disputes regarding alternative scientific and technological paths toward such securing, these disputes centering on questions such as the varied environmental costs of the several paths.

There is a further consideration. Should humanity transition to a posthuman existence, it is possible that other forms of life will be favored over the scientific and technologically cognitive mode of life that has been indicated above. For example, the aesthetic life might be favored, with the understanding that new forms of aesthetic expression and appreciation may emerge. In response to the challenge that such future aesthetic life is founded on the material surplus and other securities provided by appropriate scientific and technological developments, it is at least apparently conceivable that posthumanity has placed such development and management in the hands of capable machinery. The existence of such machinery would allow posthumans to emphasize other modes of living than those of scientific study.

In the following chapters there will be further consideration of the overall question of the manner of life for transhumans and posthumans. For now, let it be said that visions of intelligent machinery performing various intellectual tasks, at least some of which having previously been performed by humans, appear sufficiently alienating as to be discredited. This is not to disallow assistance from increasingly capable machinery. But at least one result of such assistance is that it frees people to perform ever more sophisticated intellectual tasks on the basis of this assistance, much as hand calculators allow those engaged in mathematically intensive activity to delegate various tedious calculations to machinery, the better to pursue further mathematical thought. The hope that machinery, in the common sense of the term, will someday perform all mathematical thought, thereby completely liberating our descendants from the supposed tedium of all such thought, arguably rests on a stultified view of human endeavor and enjoyment. Nor will the question be closed by noting that at issue is posthuman life, rather than human life. Our only means of projecting such life is from our present standpoint, and from that standpoint it remains debatable that we would wish to discontinue such intellectual activity.

There is much imaginative freedom as to what we might become. Not all possibilities are self-preserving, in any sense resembling our present, commonsense intuitions of selfhood. Of course, in the spirit of this work, our commonsense intuitions are subject to critical investigation. Still, there are suggestions at some remove from any account of selfhood as currently envisaged, whether intuitively or critically. Among such suggestions: we might evolve into some sort of single entity of perhaps planetary size or larger, with various capacities generally inconceivable to us now; we might keep some semblance of selfhood, yet have immediate access to other minds (all or some), however the term "mind" is understood; we might regularly reconstitute our physical natures to the extent that continuity of selfhood cannot be underwritten, to whatever extent it currently is, by intuitive bodily continuity.

Much has been written regarding personal identity. Much has been contested. Within the monotheistic tradition, the notion of a special sort of bearer of personal identity, often eternal personal identity, is familiar. I shall not pursue that notion in this discussion, because I believe it runs counter to our current and foreseeable scientific theories and results. Those in disagreement with this stance sometimes urge the compatibility of scientific and monotheistic-theological outlooks. I do not accept such compatibility regarding the current concern, but I shall not contest the point here. Those who believe in the priority of theological considerations of the sort relevant to this discussion, and not merely the compatibility of such considerations with scientific thought, are likely in disagreement with the overall point of this work. What I can offer is the following. Perhaps the eternity of self monotheistic theology posits is best understood as something capable of physical realization. My suggestion is that whatever earthly personal identity currently exists is contemplated as extended indefinitely. I do not suppose those believing in eternally existing souls will accept this offering, but I have not attempted to present it in the best possible light. Perhaps those willing to make the effort can affect something of a reconciliation.

In the context of the overall discussion of this book, there are loose constraints placed on speculative proposals regarding selfhood and personal identity over time. These constraints do function, however broad they are. Given the appeal in the preceding chapter to pointers from the present to any future we might envision for ourselves or our descendants, there are proposals which are apparently unacceptable as near-term possibilities. Included among unacceptable proposals is the above suggestion of a single, perhaps planetary-wide self. This suggestion should not be confused with what is perhaps a more familiar idea, that we evolve into entities that can access each other's consciousness so as to think at times as an intimately "wired-up" community.

This latter scheme still entails a plurality of individual selves, however much the notion of self is extended by such intimate sharing of thoughts.

The foregoing is admittedly vague and speculative. My supposition is that there is an indistinct line dividing suggestions of future beings to which we have evolved into those suggestions humanity is willing to consider, and those suggestions humanity presently is not willing to consider. I believe it unlikely that humanity, even humanity brought universally to a position of education and material comfort similar to that currently enjoyed by a first-world professional, would choose the posthuman goal of a planetary-wide self. This invites the response that it is not likely that humanity, even if universally developed as speculated, would choose any posthuman goal. I grant this response, but I nevertheless want to suppose that humanity will evolve at some future time, and that such evolution is possibly largely under the control of some portion of humanity itself. This supposition will be explored in various ways in following chapters. For now, I am making an idealized assumption that the greater part of humanity is in a more literate and generally informed state than is currently the case, and for whatever reason has decided to proceed to posthumanity. This assumption allowed, I believe there are such constraints on their choices as I have briefly indicated.

Philosophers have said much regarding the notion of self. In recent decades, various analytic philosophers have focused on the related concepts of selfhood, personhood, and personal identity over time. Such work cannot be reviewed in either breadth or depth here.[6] Still, there is a remark I want to make regarding some of the discussion, or perhaps regarding attitudes that might be suggested to some as a result of such discussion. The attitude to which I refer is that the self, being a somewhat flimsy, quasi-logical construct, is not real.

Identity of a changing physical object over time is somewhat of a pragmatic decision, at times subject to legalities or other interests. Identity of an institutional object over time, such as an organization, has its loose and diverse criteria. Identity of a person over time brings further considerations, inasmuch as a person is not only an "outward" physical configuration, but also a psychologically coherent something. Given its close relation to personal identity, selfhood, in its initial ordinary but non-religious aspects, appears to incline more to the psychological than the overtly physical.

Within this overall context of related issues, it has not proved difficult to raise interesting real and hypothetical cases which function as counterexamples to naive or intuitive notions of personal identity and selfhood. One's self (a likely redundant phrase) is less necessarily integral, or coherent, or continuant than one might commonly suppose. But what do such philosophical complexities license? Can we say, for example, that Kant is wrong in his overall point that one's making sense of their perceptual experience demands an integrated perspective, a unity of apperception? Or, if Kant is speaking of a moment-by-moment unity, does the Kantian assumption of "career continuity" of various objects of experience, in conjunction with a moment-by-moment personal perspective, necessitate a unique and continuant self?

I would not stake much on my understanding of Kant. Nor would I stake the future of humanity on Kant's having gotten the metaphysical implications of his epistemology generally correct, or on his having gotten epistemology generally correct. Commentary both on Kant and on the issues whose authorship I am ascribing to him is extensive, and I shall not pursue these topics further. However, there is a point here of relevance to our transhumanist considerations. Philosophical concerns regarding the legitimacy of our common conception of selfhood typically involve fictional suppositions such as brain-transplants, uploaded mental contents subsequently downloaded into other bodies with "swept" brains, and so forth. Such thought-experiment materials aid the philosophical mission of casting doubt on the tenability of these common conceptions of personal identity and selfhood. But what is not always pursued in these discussions is the psychological and epistemological consequences of taking the creations of such thought-experiments seriously.

Augustine noted the impossible consequences of a consciousness confined to the vanishing "present moment," and concluded that human psychological being demanded both memory and expectation. The protagonist in the film *Memento*[7] suffers from anterograde amnesia, which prevents him from keeping memories of ongoing experiences beyond short-term. Were this condition projected onto the whole of humanity the result would likely be species-obliterating. Granted, the overriding subject of this book is trans-to-posthuman selfhood rather than current conceptions of the self. Nevertheless, regarding the transitioning, it is unlikely that humanity will accept as a first step the sort of radical break with their current psychological dispositions as evinced by some of the above-referenced thought experiments.

What I have been urging is that transitioning from the human to the posthuman involves an initial respecting of various constraints, which can only be broadly indicated here. Allow an indistinct division of those constraints into those that are psychological and those that are epistemological. Then, for example, the idea that humanity becomes a single, unified entity, minimally violates intuitive psychological constraints regarding what sort of mental life individual humans are likely to desire for themselves or their descendants, whichever makes the initial, radical transition to entities which can be agreed to be no longer human. Or, the idea that one person—in the philosophically vague, intuitive sense of "person"—can "load" himself minute-by-minute into different and widely separate bodies, appears to violate epistemological constraints having to do with continuity of perceptual experience. The first example arguably also raises epistemological difficulties, and the second likely raises psychological difficulties as well. I emphasized the psychological difficulties of the first example, and the epistemological difficulties of the second, for reasons of prominence.

I do not mean to suggest that these loosely specified constraints will remain in place for all time. If speculation into the far future (say, 10,000 years or more hence) be allowed, what might at that time be the physiological, psychological, and sociological circumstances for anything that could be considered our descendant(s) will probably transcend any presently specifiable constraints, even as various current physical limitations, such as the speed of light, might be overcome in a manner we are not now able to conceive. This thought will be pursued further in later chapters.

One might challenge my apparent commonsensical attitude towards selfhood in comparison with my proposed rejection in the preceding chapter of institutionalized retribution, on the assumption of the falsehood of the ascription of what I there termed the "metaphysical" sense of freewill. By parity of treatment, why not abandon the commonsense notion of selfhood for reason of its unsustainable metaphysical foundation?

There are responses to this challenge. While I am not certain that humans lack freewill (in the sense of "freewill" of which I assume, without argument, that freewill is not compatible with either determinism or randomness), I am convinced that if humans lack freewill institutionalized retribution is not rationally justifiable, inasmuch as such retribution posits freewill as a necessary condition of its application. On the other hand, I make no such assumptions about the metaphysical character of selfhood. It may be that self is unreal in the manner that I believe free will is unreal, but I do not possess adequate reason to assert that ordinary cases of selfhood are any less real than the solid-seeming, phenomenally colored table that Eddington contemplated.[8] Nor do I have in mind any hypothetical of the form, "if the self is not real then . . . ," in the manner in which I believe institutionalized retribution is untenable in the absence of freewill. Further, I do think we would be a better human society if we abandoned our faith in freewill, and concomitant judgments of moral responsibility (although not causal responsibility), although this is a belief I cannot defend here. I do not think any present or near-future (i.e., possibly transhuman) purpose is served, however, by abandoning some non-theological common ideas of selfhood.

I have proposed three metaphysical circumstances be preserved in our transition from the human to the posthuman.

First, we should retain consciousness, by which I mean something more than awareness or alertness. I believe it is possible for an entity to be aware and alert, yet not fully conscious—if conscious at all. Anyone who has tried to eliminate a cockroach is likely impressed with the insect's level of alertness. Whether a cockroach is sufficiently ganglionated to have the sort of consciousness capable of feeling pain, experiencing dreams, or being joyful (in more than the behavioral sense—although likely not even that for the cockroach) is problematic. If one wants to allow such for the cockroach, then consider the level of alertness a paramecium displays under microscopic examination. It is difficult to ascribe the sort of consciousness indicated above to this apparently in some sense alert, unicellular creature.

The earlier hopes for artificial intelligence (AI) were centered on computational machinery that could perform various cognitive tasks. In his famous paper, Turing[9] argued that any such machine capable of a no-holds-barred conversation with a human being could justly be pronounced intelligent. Agreed, but sentience, or consciousness, is another matter. Unfortunately, at times AI, including robotically embodied, connectionist enabled, being-in-the-world AI is entertained as a possible evolutionary goal for humanity, without concern for the sort of consciousness of which I speak. Granted, there are those who defend the position that such AI is all that is the case with consciousness. I disagree, and hope that if and when humanity takes its first steps beyond itself, the consciousness to which I refer is on board.

Second, the world with which transhumans are in primary, cognitive engagement is understood to be the real, external world. I believe it serves philosophical purposes best that such a world be understood to be something more than a logical construction of sense data, or an indirectly sensed cause of sensation (with a degree of resemblance to sensations ranging, in various accounts, from zero to full), but I am especially aware of the controversies here, which may never be settled with anything similar to the near-unanimity that attends various well-entrenched natural-scientific positions. I content myself with the following. However one philosophically apprehends the phrase "external reality," there is enough commonly intended to prohibit such post-evolutionary goals as living in a computer-simulated, virtual reality. However convincing the simulation, however communal the involvement, I cannot accept that humanity would choose to be primarily occupied with such an environment, rather than having primary affective and cognitive involvement with what I have been terming "external reality."

Third, if and when we transition to posthumanity, we will arrive initially as integral selves, much as we currently consider ourselves. While there are other possibilities regarding selfhood, some of which dispense with the self as it exists in common currency, the realization of any such possibilities will await such time in the future as our posthuman descendants can approach them with confident understanding. I believe that we do not now possess such understanding, and that we are unlikely to possess it in what I speculate will be our initial transitioning.

I take metaphysics to be in some part the study of extensive, possibly ultimate, constraints. Letting "ultimate constraints" mean "constraints that cannot be violated under any conceivable circumstances" (itself a phrase that warrants elaboration), I do not know if there are any such constraints. Quine might be correct (in "Two Dogmas")[10] in his belief that there are none. Some philosophers would rather speak of epistemological constraints than metaphysical constraints. I see these two philosophical approaches as inextricably linked (is my use of "conceivable" above metaphysical or epistemological?) and am indicating an indefensible preference by speaking of the metaphysical, rather than the epistemological. In any case, those constraints that are not ultimate have what I consider varying degrees of hardness (or softness). The three constraints I have emphasized are mostly soft. I can imagine human, or human-like, mentation without dualistic consciousness. I can imagine the worst regarding the external world; namely, that there is none, or none with which I am in contact. I can follow Hume in considering the self as being inwardly nothing but an associated play of ideas, and outwardly a questionable continuant.

In the preceding chapter I spoke of pointers from the present to the future. I hope that any considered choice we make as a species towards a posthuman future is influenced by careful thought as to what is best. I believe that from our present standpoint, the constraints I have reviewed in this chapter will be considered appropriate guides to our chosen future. But they are soft constraints, and I can imagine conditions under which we, or our successors, might choose to violate them. For example, if external reality were to become in some manner unbearably horrific, we might choose to mitigate the horror, if we are able, by primarily associating within a virtual reality of our creation. I leave it to the reader to imagine analogous violations of the other constraints.

In the fullness of time, of course, there may emerge currently unimaginable considerations, either of avoidance or of attraction, for violating these constraints. For now, I believe they stand.

NOTES

1. There are far too many works to recommend. If the reader desires an introduction by way of competing views, my recommendation is the following two works: Daniel Dennett, *Consciousness Explained* (Boston: Little, Brown and Company, 1991); David Chalmers, *The Conscious Mind: In Search of a Fundamental Theory* (New York: Oxford University Press, 1996).

2. The Work of Benjamin Libet arguably allows for a more direct dating of what I am terming "epiphenomena." See his "Do We Have Free Will?" *Journal of Consciousness Studies* 6 (1999): 47–57.

3. Rene Descartes, *Meditations on First Philosophy*, trans. Elizabeth Haldane and G. Ross, *The Philosophical Works of Descartes* Vol. I (New York: Dover, 1931).

4. A good example of such discussion is found in the Spindel Conference on Vagueness, *The Southern Journal of Philosophy* 33 Supplement (1994).

5. J. Austin, *Sense and Sensibilia* (New York: Oxford University Press, 1962).

6. Kept to one book, I recommend Derek Parfit, *Reasons and Persons* (New York: Oxford University Press, 1984).

7. Released by Newmarket (2000).

8. Arthur Eddington, *The Nature of the Physical World* (New York: Macmillan, 1928). In this work, Eddington worries which of two tables, the one of ordinary perceptual experience, or the "scientific" table which "underlies" the former, is the real table.

9. Alan Turing, "Computing Machinery and Intelligence." *Mind* 59 (1950): 433–60.

10. Willard Quine, "Two Dogmas of Empiricism," reprinted in his *From a Logical Point of View* (Cambridge: Harvard University Press, 1953).

Chapter 3

On Transition

> "We owe it to the brain of our species to stand here and die, without destroying a source of wisdom." (*The Thing from Another World*, RKO Pictures)

The phase of transition from the human to the posthuman demands focus. It is one thing to speculate on the joys of a posthuman existence. It is quite another to think soberly about getting to that joyful state. There have been few large-scale revolutions in human history that have been unaccompanied by large-scale suffering. Even those grand changes that, from a political perspective, are accorded the epithet "revolutionary" metaphorically, such as the industrial revolution or the scientific revolution(s), have often been attended with severe dislocations of individuals and groups.

In this chapter, I want to focus on the socio-political aspects of my speculated movement from the human to the posthuman. Although what is said here will be cloaked in talk of probabilities, the obvious note should be sounded that, as with all else in this work, the nature of this discussion is speculative. Intuitively, there are rough degrees of speculation. Talking about what is to happen the following day in human life is less speculative than talking about what is to happen a decade hence, but it is speculative. Much unexpected intervention is possible between today and tomorrow. This noted, the following discussion will seem comparatively moored in recognizable realities when held against much else in this work, and in related work of mine. As such, the following discussion is more open to commonplace critical scrutiny.

To temper the somber mood of the ensuing investigation, I begin with an explanation of the above quotation. The quotation is from a classic science-fiction thriller. The speaker is a sleep-deprived chief scientist at a North Pole scientific outpost, circa 1950. A bloodthirsty extraterrestrial threatens the inhabitants, scientists, and air force personnel. The chief scientist wants to communicate with the alien to acquire as much of its knowledge as possible. The military men and most of the scientists want to destroy it; it constitutes a threat to humanity in its ability to reproduce quickly and its malevolent intentions. The overall tone of the film evokes no sympathy for the position of the chief scientist. The alien is powerful, hostile, cunning, and vocalizes in the growls of a puma. The lighting is black and white, with much shadow, and the score is in tones of dread, appropriately laced with theremin motifs.

It is likely that no one watching the film would choose any option other than fighting the alien. Yet most watching the film respect the sort of knowledge the alien possesses. Divorced from the frightening alien, powerful mindfulness coupled with advanced scientific and technological knowledge are items many humans in developed societies would welcome, if bestowed as gifts. Doubtless there would be qualifications and caveats. One's limited lifespan lessens one's enjoyment of such ability and knowledge (the alien is not subject to animal-like mortality). It is probable that possession of such abilities conflicts with other facets of human life that few would willingly relinquish. In any case, it is unlikely that many humans would follow the quoted suggestion of the chief scientist. Let this suggestion be termed the "first scenario."

We can imagine an altered scenario. Benevolent extraterrestrials visit, bearing material and intellectual gifts. Humans learn much from them, but cannot match the visitors' cognitive abilities. Nor can individual humans outlive them, for they have indefinitely extended lifespans. At some point in the amicable alien-human interaction the aliens make a startling offer which the humans are free to reject. The aliens have grown fond of planet Earth and would like to inhabit it. However, for whatever reasons, they cannot fully inhabit the planet as long as humans—in anything like their present numbers—inhabit the planet. The aliens propose that the human race largely discontinue after the present generations live out their years—in complete comfort courtesy of assistance from the aliens—and the aliens then take control of the Earth.

I venture that many humans, if not most, would reject this offer. I lack the means to confirm this intuition, however. I also cannot confirm my intuition that, to most listeners, this second scenario is more palatable, or at least less unpalatable, than the first.

Consider a third possibility. As human science and technology progress, the means become available greatly to enhance human cognitive ability, as well as to extend human lifespan indefinitely. Assume these improvements or, if preferred, alterations necessitate various changes in human physiology and psychology, changes which will alter human social relations. There will be other effects. For example, if humanity confers on itself indefinitely extended life, there will be scant opportunity for children on whom such will also be conferred. At most, planet Earth could tolerate only one or two such additional generations. In short, in choosing such alterations humans will be choosing not to remain physiologically, psychologically, or sociologically human in commonly accepted senses of the term. Allow one further assumption: if and when the choice so to evolve is made, those making the choice do not—for whatever reason—receive the benefit of the choice, which can only be conferred on some future generation.

Plumbing my intuitions again, I find that I am uncertain as to the reception this last scenario would receive. Some might like it, some might not, and some might be uncertain. Many would likely want a more detailed description prior to voicing an opinion. This desire for detail contrasts with what I presume would be a rejection, on hearing, of the first two scenarios, neither of which are given in detail (the airmen triumph in *The Thing*, and nothing is presented here of either the horrific alien's plans for the planet, or those of the benevolent aliens, once the humans are removed).

The reader will have gathered that these three scenarios have similar outcomes: in each a life form other than homo sapiens governs the Earth. In each scenario the dominant life form is cognitively advanced compared to homo sapiens, with an indefinitely extended lifespan in which to enjoy its cognitive powers. Let us further assume that these enhanced cognitive powers have bestowed appropriate controls on individual behavior, so that members of these species live cooperatively with one another, and do not endanger or significantly diminish one another's lives.

There are nevertheless recognizable differences which legitimize the different resultant attitudes on the part of those hypothetically surveyed. Humans do not wish to die violently, and such death is what the alien(s) of the first scenario portend. Humans wish to be in some control of both their own future, as well as that of succeeding generations, and the latter such control is removed in the benevolent alien scenario.

The preceding discussion is intended to heighten appreciation of the following question. Given that we, homo sapiens, are to transform ourselves (or be transformed) into a species other than we are, into what are we to transform ourselves and how, and to what degree, are we to manage this transition (or this series of transitions)? This compound question is obviously restricted at this time to speculative answers. Still, the freedom inherent in speculative activity can be subject to intuitive limitations.

I want to focus on one aspect of the above question; that of the manner of management of the transition. Reflecting on the preceding scenarios inclines me to believe that we desire some control over such transition. Further reflection indicates three critical variables: who we are, the nature of the control, and the nature of the transition. With reference to the first variable, I do not say "the nature of us" because this is not the whole of the question. If we are human, then the problematic intended by "who we are" is the question of which humans have control of the transition (i.e., the emphasis is on "we"). Is it all humans or some, and if the latter, which? Since it is not expected that infants and young children are included in the controlling group, "all" is taken in whatever intuitively reasonable sense can be made of "adult human." But "we" might not all be human. Some of us might have transitioned prior to the question of control being raised as a matter of public concern. This complication will receive attention below.

A detailed, albeit speculative, investigation of these three variables will be complex. This is partially due to the complexity of each of the variables, some of which are indicated above in the consideration of the first variable. Additionally, these variables interact. For example, assume some but not all humans have transitioned to posthumans, and these posthumans have significantly advanced cognitive abilities in comparison to humans. This assumed fact is relevant to the question of control of the transitioning of the remaining humans. I do not say here that it is in the interest of the remaining humans to relinquish control of the transitioning to these posthumans. Depending on further factors, the contrary might be advisable. But this assumed large difference in cognitive powers will likely need to be taken into account.

Let us explore the terrain over which these variables range. Suppose we allow the following idealized assumption. The human condition worldwide is optimized to the extent that virtually everyone lives at the highest material and intellectual standards such as are currently enjoyed by the middle class of the most prosperous countries (for example, Denmark). This assumption is both vague and debatable. What are the details of such material and intellectual standards? Why suppose these details are realized anywhere, let alone by the middle class in Denmark? Philosophers who have occasionally favored some forms of aristocratic life, such as Plato and Nietzsche, are easily imagined to differ with the positive value implicitly imputed in my presentation. Am I assuming that none of these middle class people are leading "lives of quiet desperation"?

I grant these and other unspoken criticisms of my idealized assumption. Given what I have written in "The Future of Ethics," I do not believe prevailing human nature allows for the realization of this assumption, even allowing for the vague and intuitive form in which it is presented. The assumption nevertheless serves as a limiting case of a range of possibilities. Let me add the following burden to this assumption. There exists an ethical theory comparable in its scope, clarity, detail, and degree of confirmation to that of well-entrenched grand theories of natural science, and this theory is accepted by humanity worldwide to roughly the same extent that educated people accept the general theories of physics. I am aware that this will be considered an outrageous assumption. I am also aware of stating in the above-referenced earlier chapter of this work that I do not believe in the possibility of such a theory, given current human nature.

If the reader is willing to continue in the face of these vague and questionable suppositions, I request the following further suppositions be allowed. From this speculated ethical theory, taken together with the speculated worldwide circumstances regarding the state of humanity, the inference may be drawn that it is in humanity's interest to transition to posthumanity. Further, the means of this transition, be they pharmacological, surgical, genetic manipulative, some combination of these, or possibly some other, are available and known to be available. Finally, it is understood that this transition can be accomplished with minimal discomfort to individuals and groups.

I may have missed something in my above assumptions. If I have omitted an important consideration, in the spirit of this idealization I am hopeful that yet additional assumptions will be corrective. It seems obvious that, given these assumptions, we should and would desire to transition to posthumanity. Of course, what is in humanity's interest as a whole might not be in every individual human's interest. We might want to add that any individual human is permitted either to remain human, or to transition back to being human from being posthuman (let us assume the means exist for this return), should they desire.

The above list of assumptions exemplifies an idealized set of circumstances in at least two senses of "idealized." Various specifics have been omitted for the purpose of simplifying the intended inference, and the assumptions made are clearly best case. There is scant reason to believe that the assumed circumstances will prevail in the real course of things. What appears more likely is a piecemeal transformation of humanity, where the transitions are neither complete in terms of desirable changes nor in terms of peaceful co-existence of the entire human and posthuman population.

In the idealized case above an inference drawn from the assumptions is that a majority of our descendants will have desired transformation. Further, the flow of the discussion suggests a one-time decision on humanity's part to accept an encompassing package of alterations. The compound assumption that this transformation will be desired by the majority of humankind at the appropriate time of decision, and that the decision will be a singular one, is problematic. It is particularly unlikely that there will be only one decision to make, although were that the case matters would be simplified. A more probable scenario, however, is one in which a series of decisions, separated by substantial temporal intervals, are to be made.

There are other manners in which the idealized assumptions are obviously too neat. Even if the appropriate scientific discoveries enabling radically extended lifespan are imagined postponed until humanity is in a settled socio-economic-political situation of relative parity, there remain vexing loose ends. A majority of billions allows the possibility of a minority of upwards of billions. Will a substantial majority of this minority consent to the will of the majority and, against their judgment, themselves undergo the transformation (or have their children undergo it, if that is the issue)? And, regarding the immediately preceding parenthetical, if the decision regards one's offspring, what are the age-limitations on liability? Will the transformation be limited to those under adulthood, those pre-pubescent, only those in infancy? Any designated cut-off not keyed to what, in some choices of such cut-off, is a currently undiscovered biologically clear-cut criterion, promises to be received as arbitrary.

Consider two profound transformations: indefinitely extended lifespan and significant increase in cognitive abilities. Although the former speculated alteration is apparently singular, the latter admits of various differing manifestations. Actually, there are also various sorts of indefinitely extended lifespan. One dimension of such variety is the degree to which an individual's lifespan is protected against external shocks. However, let us make the simplifying assumption that the opportunity for increase of cognitive powers, as well as the opportunity for indefinitely extended lifespan, each becomes available as a singular package. Allowing this simplification, it remains that there is a plurality of alterations under consideration. These two considered alterations are of sufficient difference, both logically and empirically, that attaining one does not appear to necessitate having the other. It may be, however, that having only one of either will more incline the possessor to want the other. Further, a significant increase in cognitive powers, appropriately manifested, heightens the probability of discovering means of securing indefinitely extended lifespan. Alternatively, humans who have been transformed into beings with indefinitely extended lifespans, all else held constant, will likely want an increase in their cognitive powers, given that many current humans would want such an increase if it were available. Researchers in relevant areas will have more time, as individuals, to focus on the problems of achieving this goal.

These two speculated transformations have been chosen because of their likelihood. It might be said that they are too likely. In particular, dramatically extended—if not indefinitely extended—lifespan, could be just around the figurative corner. Overcoming the various well known limitations on cellular replication, regeneration of nerve tissue, organ replacement (natural and artificial), are subjects of active research, rather than merely fanciful imagination.[1] In whatever form such extended lifespan takes, there will always exist the possibility of life-terminating small-scale or large-scale cataclysmic events. Unexpected individual traumas, such as being incinerated in a lightening flash, are ever-present possibilities. Barring such events, however, prospects for dramatically extended lifespan are bright. In the far-off future the means for recovering from at least some such life-destroying events might be at hand. For example, individuals might exist with simultaneous "back-up" systems which, given then-existing technology, could replicate them whole in their physical and psychological aspects.

If the assumption of a relevantly equalized world population is revoked, and the somewhat likely circumstance transpires that relevant scientific discoveries are made in the near future, then widespread social conflict is to be expected. Well-positioned people in the developed sector will likely be the first to choose to transform themselves and/or their offspring. Many of the others, with varying degrees of lack of access to scientific and technological discoveries, might be positioned as resentful observers to those considered the fortunate few. Of course, if the transformative means were available to all, as is the case with various vaccines, lack of access would not create special difficulties. Still, cultural and economic differences would likely influence choices, and attendant resentment of those choosing otherwise. The interim of alienation will manifest sufferings that would not exist but for the presence of the means of transformation. Further, various upheavals will continue for a period after the majority of humanity chooses the transformation, such upheavals being in some measure the result of many wanting to continue the practice of having children.

The near-term forecast is hence not fully pleasant. The formerly human population is now considered as consisting of those with radically extended lifespans, and those with ordinary human lifespans. Given speculative liberty, we might conceive the former either as physiologically and psychologically similar to the ordinary humans, or to varying extent not so. These conceptions are significantly apart. In the first case much social and sexual congress is likely, with hitherto unencountered problems arising as to the conduct of the union between the altered person and the ordinary human, assuming a union of some duration. There are further difficulties regarding the nature and treatment of any offspring of such union.

Regarding the case in which there are marked differences between those with extended lifespans and ordinary humans, some segregation of these different populations is to be expected. Such separation might be geographical, or social, or political, or some combination of these. Even supposing these forms of separation are minimized, it is difficult to imagine a lack of psychological distancing of these two groups. Suppose, for example, that the posthumans are physiologically sexless. Or, that they have other different physical capabilities and constitutions than their human counterparts. For instance, they might have dissimilar sleep requirements, or sleep rhythms, than ordinary humans. Any or all such differences promise a psychological divide between the now varied species.

There are more variations on the above themes than can be investigated here. Another significant variable is the comparative size of these two populations at any given time. Another is the comparative wealth of the two groups, assuming some standard of wealth can be found. Yet another significant question is the distribution, or lack thereof, of political power, with the attendant question as to how such power is manifested (consider, for example, various durational matters such as those regarding terms of office). Some measures of value will vary between the two populations, even supposing they are physiologically similar. A month's vacation is likely to be more prized by an ordinary human than by someone with an indefinitely extended lifespan. Assuming members of these two groups could be similarly employed, retirement will be a different matter for the humans and the posthumans.

All such difficulties might be regarded as resulting from an initial fanciful speculation; that some descendants of humans become something other than human. These difficulties are avoided on the assumption that no such change occurs. Give current research efforts in relevant areas, however, arguably the more fanciful assumption is that no such transformational circumstances happen. Discussions of this sort benefit from the occasional sobering reminder that extended lifespan is a prima facie possibility, and that such extension might not initially be available to all segments of the population, nor might all people willingly accept it even if it is available to them.

Those unwilling to countenance a transition from the human to the lifespan-altered posthuman can resist the suggestion of such alteration intellectually, politically, or forcefully. These three general modes of resistance are understood here in a nonexclusive sense. Focusing on intellectual resistance, the following forms come to mind: the conceivability of such transition might be challenged; such transition might face intractable difficulties of the technical-scientific sort; ethical challenges might prove insurmountable (this last intellectual difficulty might especially fuel successful political resistance to the transitioning at issue).

The existence of a clear delineation of the conceivable from the inconceivable has proven philosophically problematic. In practice, what has been considered conceivable has ranged from the empirically imaginable (itself a problematic criterion) to the formal-logically consistent. Matters here are not as anarchical as might initially seem. If a putative state-of-affairs is formally inconsistent it is likely unimaginable. Of course, the converse only holds for the more hardline versions of empiricism. In any case, assuming a tolerant stance on the matter of the meaning of "conceivable" allows for a hearing of the sort of claim that might be made to the effect that transitioning from human lifespans to indefinitely extended lifespans is inconceivable. If this claim can be established, or even granted a strong probability, so much the worse for any considered posthumanism that relies on such extended lifespan. Alternatively, if this claim of inconceivability is not established, then depending on the fullness of the investigation, the charge of inconceivability might be eliminated or marginalized in these discussions.

Various fictions premise human or human-like characters who, but for their extended longevity and, perhaps, hidden differentiating quirks, are otherwise leading mostly ordinary human lives. Vampires might have special needs, but in some fictional venues they get along well with the surrounding humanfolk. What is less readily found in science fiction and related genres are scenarios of significantly large populations of descendants of humans with indefinitely extended lifespans coexisting with ordinary humans. Those fictitious lone or few "immortals" that occur more readily in these fictional venues function in this discussion as a thought experiment, a test of the conceivability of such characters in otherwise realistic contexts. If coherent, detailed dramatic representation constitutes a successful result, then we might want to conclude there is no conceptual difficulty in considering a limited population of humanlike individuals whose durational differences with ordinary humans remains hidden. We might be tempted to conclude further that the sometime success of such fictitious "immortals" at remaining undiscovered while in the midst of ordinary humans offers evidence of the factual possibility of such existence. At the least, these familiar and widely-disseminated fictions tend to incline us to accept such possibilities uncritically.

Approaching these fictional representations with a critical attitude, however, we note other often represented themes, such as time travel. In particular, the theme of time travel to the past for the purpose of altering selected features of the present—including the time traveler, is a staple of recent science-fiction cinema. Under critical scrutiny, such time travel has an air of logical impossibility or, at the least, great factual difficulty. Perhaps there is some similarity between the case of time travel and the case of immortality amongst the mortals. It seems possible, even likely, that difficulties are being overlooked. Various difficulties have been mentioned above, but for the most part they arise on the assumption of marked physiological differences between ordinary and transformed humans. Even if the conceivability of an occasional, lone, human-appearing individual living an indefinitely extended life among humans is allowed, there is a question more relevant to this inquiry: is it conceivable that a substantial group of posthumans be living together with ordinary humans?

Even as this question is asked, the answer is apparent. This does not seem a matter of conceivability so much as a matter of factual difficulty. If there is any conceptual difficulty in the supposition that humans and posthumans live in close proximity, such difficulty can be ameliorated by considering the two groups at some remove from one another. That there are factual difficulties cannot be denied; various such have been mentioned. What is of concern now is the extent to which the possibility of indefinitely extended life is supported by current natural-scientific theory and research. To return to science-fiction for a moment, the notion of teleportation featured in *Star Trek* venues has been challenged as physically untenable.[2] Perhaps from sober-minded biological science there arise similar objections to the feasibility of indefinitely extended lifespan.

I have been stating that extended lifespan, even indefinitely extended lifespan, might be realized in the near future. Scientific opinion is apparently divided on this possibility, at least as far as the topic of aging is under consideration. There are minimally three impediments to indefinitely extended human lifespan (or "healthspan"): aging, disease, and environmental impacts. As human life is presently constituted, there are various life-threatening diseases whose probability increase as a person ages; perhaps most notoriously, cancer and heart/circulatory disease. The question of when all such life-threatening diseases will be eliminated, or significantly limited in occurrence, is beyond a confident answer at this time. That such elimination will happen in the next several decades is extremely improbable. Still, if the most notable of these diseases, cancer and heart/circulatory disease, have strong links to aging, then the retardation/reversal of aging might lower the risk of contracting these afflictions.

As long as we humans remain in anything like our present biological form, we will be subject to termination via environmental impact, whether at the hands of other humans, or through nonhuman causes. One speculated manner of response to this general threat has been mentioned above; that means exist for "backing up" our psychological aspects and "downloading" them into convenient biological receptacles, such as cloned replicas of our former physiological selves. The probability of near-term implementation of such schemes is unlikely. Of course, another manner of avoiding, or at least mitigating, external threats is to transform ourselves into radically different physical entities, perhaps largely or completely nonbiological. This manner of avoidance has the added benefit of potentially avoiding many of the problems of disease. While such alteration might be a far-distant mode of evolved humanity, it has little prospect of realization in the near future.

If indefinitely extended lifespan is to be realized within the twenty-first century, it will almost certainly be due to progress on retardation, and possibly reversal, of aging. This is so, given the unlikelihood of overcoming all life-threatening diseases and environmental impacts. It is possible that progress on extending lifespan will be incremental in two senses: initial extensions might be limited (e.g., an average of two hundred years), rather than indefinite, and the percent of worldwide population receiving such extension might be limited initially, and perhaps indefinitely into the future. In the case of limited duration lifespan extensions, one important variable is the extent to which those having limited extensions can be "upgraded" or "retrofitted" so that their lifespans are further extended. In the case of limited distribution of extended lifespans among the initially human population, perhaps the most significant variable is availability, although another notable variable is willingness. The effect of such variables, assuming they do not acquire limiting case values, is a mixed population of ordinary humans, humans in transition, and posthumans. This result will be examined further below.

The biological phenomena of aging, as manifested in complex animal organisms such as humans, is not well understood in all its aspects, and theories are to some extent in competition.[3] However, there is much that is known, including the limits of cellular replication, the collection of by-products of cellular metabolism and cellular decay, and various forms of tissue and organ damage. While even greater understanding would be useful in the attempt to retard and reverse processes of aging, such understanding is neither necessary nor sufficient for progress. There may be a soon-to-be-discovered "magic bullet" that extends lifespan, although for ill-understood reasons. Whatever progress is made on biological fronts, there is and will likely continue to be research on protheses. How far such work on artificial replacements will go is a topic of speculation ranging from the scientifically and technologically informed to the fanciful. The sort of artificial heart having an onboard power supply, low failure rate, and sensitivity to neural and glandular input, is a subject of current research. A fully artificial brain, on the other hand, is at this time an object of idle fancy, as is the related thought of computer storage (with frequent updating) of a living person's mind (or brain-configuration). In the context of this discussion, these are important limitations. An indefinitely extended lifespan that does not include appropriate modifications of current "brainware" is likely to curtail cognitive operations at some future point of the beneficiary's indefinitely extended lifespan. We may hope, even expect, that as our understanding progresses such limitations will be overcome. For now, life extensions on average of, say, one hundred years, will not be affected by such limitations.

In between scientific-technological likelihood and wishful thinking are various non-prosthetic schemes such as gene therapy over a broad spectrum of genetic disorders and full-range tissue and organ repair either by means of stem cell introduction, or growth of tissue or organ in nutrient bath followed by full transplantation into the body (therapeutic cloning).

In sum, while there is much to be discovered, or shown undiscoverable, purely scientific and technological considerations tend to support the position that, with regard to dramatic—if not indefinite—extension of human lifespan, there is reason to believe that the question is less "if" rather than "when."

We now consider questions regarding extended lifespan that arise from ethical considerations. While I have been defending the possibility of significantly extended lifespan against objections based on inconceivability and scientific-technological impossibility, I am less sanguine in the face of moral objections. Given these objections, my support of the endeavor of various humans to transform themselves or their posterity into beings with radically, ultimately indefinitely, extended lifespan is at best qualified.

Prior to an examination of ethical concerns I want to note that, were the idealized assumptions discussed above implemented, I would have no reservation regarding the transition of humanity to beings having indefinitely extended lifespan. Further, my endorsement of this transition does not require the full implementation of these idealized preconditions. The fulfillment of one condition that was not part of the list of idealized assumptions above would, other matters being roughly as they now are, result in my consent. That condition is that a significant number of humans, or perhaps humans in a transitional state, be endowed with cognitively superior minds.

The most immediate problem with the phrase "cognitively superior minds" is the satisfactory specification of its intention. A second pressing problem concerns the relation of cognitive abilities to moral thought and behavior. I have addressed these problems in my book, *Inhuman Thoughts*, particularly in the chapter "On Human Improvement," and much of what follows is drawn from this previous work.

I want to limit my elaboration of what is intended by the notion of cognitively superior mindfulness because I am not able to supply a thoroughgoing account of what a cognitively superior intellect would be, such that this account would satisfy all inquiries. To say all that I want to say regarding the conception of such intellect would require excessive length, and would not improve significantly my treatment of the problem of the relation of such intellect to the moral life.

I choose a quotation from my earlier work to suggest what I mean by a "cognitively superior mind."

> Consider the expectation that children before adolescence be literate in science to the extent that organic chemistry, relativity and quantum theory, molecular biology, and so forth are understood by them to such degree that they can routinely solve the sort of exercise problems currently given in texts read by graduate students specializing in these various disciplines. That is, the average child manages this for all these disciplines, as the average child today manages adequate performance in arithmetic, basic reading skills, basic knowledge of national history, geography and so forth.[4]

In stating this as indicative of cognitively superior mindfulness I do not mean to limit the application of this mindfulness to natural scientific thought. Rather, I am thinking of minds that evince performance in all cognitive areas far in excess of current average minds, or indeed any past or current human minds. Although I speak of children in the above quote, my thought is that such children extend their cognitive powers as they mature.

What remains at issue is the extent to which minds such as those indicated above are, by their developed nature, other-regarding rather than selfish. Put to purposes that are morally questionable, such minds would be capable of great harm. Admittedly, I am assuming that terms such as "harm," "selfishness," and many other value-laden words and phrases designate possible states-of-affairs, and are not reducible to strictly emotive content, nor are they essentially non-cognitive. Further, I am assuming that in their designation of states-of-affairs, these terms are not to be considered relative in the sense in which that which is relative is contrasted with that which is constant. I am aware that further elaboration is wanted, at least regarding the last point, and I am aware that I have already stated enough to provoke disagreement. Given the magnitude of what is involved here, I must be content to speak to those who either are willing to accept this vagueness because they incline towards what I seem to be indicating, or who are minimally willing to suspend judgment in the interest of hearing what follows. This disjunctive class will not include everyone.

I am not confident that beings evolved from humans, and possessing cognitively superior minds, will ipso facto behave to high moral standards in matters of supporting the general welfare when conflicts exist between the general welfare and perceived personal benefit. Among the relevant variables are social environment and early training. While these factors might not affect those having enhanced cognitive abilities in the same statistical manner in which they effect ordinary humans, it is unlikely that they will have no effect on these cognitively enhanced beings. On the other hand, I incline to the belief that if an appropriate range of circumstances are present, those evolved beings with cognitively enhanced abilities will be superiorly gifted moral agents.

Let us suppose that those beings with enhanced cognitive abilities are morally inclined. Such inclination might be instilled at their inception, or it might be imparted by instruction, or it might result as a combination of these two sorts of factors. Behaving morally, having appropriate regard for others, is in some measure a matter of habit, and in some manner the result of deliberative thought. To the extent that moral behavior is a matter of habit, we should hope that those with enhanced cognitive abilities will develop better understanding of the means for imparting habits. That such understanding, and the implementation of what is understood, will transpire is not guaranteed by the supposition of enhanced cognition. The various forms of training that are practiced today are frequently subject to study and modification. It is possible that such study, performed by those with enhanced cognitive abilities, will produce results that are superior to current practices.

My apparent confidence in the likelihood that these beings with enhanced cognitive abilities will lead exemplary moral lives is of course based on the assumption that moral behavior is to some extent a function of cognitive ability. A cursory examination of a group of moral theorists, Plato, Kant, and Mill, lends support to this assumption. Plato's oft-repeated belief that those who know the right manner of behaving (and, in particular, know that such behavior is in their interest) will behave properly needs no further examination in the limited context of this discussion. Kant's theory of moral behavior that such behavior must spring from a will committed to the rational lawfulness determined by a certain form of reasoning, explicitly connects morality to cognitive ability. Mill, however, appears somewhat the exception. In considering the extent to which an agent need consider the effects of that agent's behavior, Mill writes:

> The great majority of good actions are intended not for the benefit of the world, but for that of individuals, of which the good of the world is made up; and the thoughts of the most virtuous man need not on these occasions travel beyond the particular persons concerned, except so far as is necessary to assure himself that in benefitting them he is not violating the rights, that is, the legitimate and authorized expectations, of any one else.[5]

While such deliberation, as Mill indicates, requires less than superior cognitive abilities, there is the disturbing aspect to his utilitarian thought that, ordinary actions having wider consequences than Mill's understandable (given his moral theory) estimation of their limits, most humans are in the position of, at best, striving to be apparently moral. However, consider the possibility that agents have the cognitive powers to take more information into account, and to reckon the probability of various wider-ranging results of alternative choices, and to do all this in realtime situations. Of course, agents might also devise appropriate aids (perhaps computational) for taking such information into account. Were such possibilities realized, Mill's theory would be less subject to the criticism that many consequences of actions are beyond an agent's ability to contemplate.

If moral theories as diverse as those of Plato, Kant, and Mill, explicitly or implicitly suppose cognitive efforts on the part of the agent (recall Plato's description of the path to genuine knowledge in the concluding pages of *Republic*, Bk VI), then there is some basis for the hope that cognitively enhanced agents will possess superior moral characteristics.

It is unfortunate that the more likely event is the availability of procedures for radically extending lifespan in the near future, without there being means of appreciable cognitive enhancement. The only manner I foresee such an event not creating worldwide upheaval is that the real cost of this procedure allows its availability to virtually all humanity in similar manner to the availability of various vaccines to most of the world's population. Granted that not everyone would avail themselves to such a procedure, those not wanting the procedure would conceivably be treated with respect and consideration. It seems more likely, however, that the real cost of the procedure radically to extend lifespan is such that, for a substantial initial period, its distribution will be significantly limited. We are not discussing predictable matters, but for the procedure to be globally available methods of delivery would require the economy and scale of various worldwide vaccinations. Such requirements eliminate the possibility of worldwide use of even the simplest surgical interventions. And, of course, it is possible that there will be no one treatment that will radically extend lifespan, but rather a series of routines, some perhaps pharmacological, some perhaps surgical (perhaps including prostheses), and perhaps some concentrating on lifestyle.

As was stated earlier, there are too many variations on the transitioning of segments of humanity to posthumanity to admit of detailed examination, even if the discussion is restricted to the dimension of radically extended lifespan. Further, the speculative manner in which any such discussion must be conducted presently will not allow us to consider the effects of the appearance of the unexpected. One need only contemplate the various technological innovations of the past century to appreciate the force of the unexpected. Add to such force the unexpected socio-political developments of the past century. All this noted, we might attempt a generalized treatment of some possibilities with the hope that such sketching can be extended and detailed so as to cover appropriate future circumstances.

If the notion of a free marketplace prevails, or minimally, if access to life extending procedures is limited to those who can afford them, then we may expect a significant disparity in the general human population. For the while, let us assume that the cognitive advancements contemplated above have not occurred, and that except for the difference in life expectancy (and whatever is necessary to effect such difference), the overall population is essentially human. In particular, the "extendeds" and the "nonextendeds" can interbreed and have fertile offspring who may or may not be among the "extendeds" at birth (this last is obviously an important difference, but I ignore it here).

A sober examination of the behavior of humankind, even one restricted to modern times, indicates the inability of diverse groups living in proximity to ignore the differences that make for separate groupings. Exceptions to this general finding are insufficient to mitigate the importance of the often violent hostility shown by these groups to one another. Attempts to blunt this fact by noting that it appears most often to involve those populations among the underdeveloped overlook recent European history, as well as various contemporary trends among Europeans and the Americas. The implicit contempt (or, if this is too strongly put, the "benign neglect") of the propertied for the economically disadvantaged, both in their midst and geographically distant, but subject to their economic colonialism (enforced by military might, if necessary), outlasts all official rhetoric of concern. Nor is there factual support for the sometimes alleged nobility of the impoverished, living in the midst of the developed societies. And the mutual fierceness shown by ethnically diverse groups of the poor whom circumstances have made neighbors is matched by recent events in the former Yugoslavia, involving the middle classes as well as those of less means.

Given such friction among various groups of humans, it is difficult to imagine fully mutual tolerance and respect of differences between the extendeds and the nonextendeds. How this will play out cannot be predicted with any degree of confidence, except to state that it is unlikely to be untroubled. In the spirit of this study, I hope that radically extended lifespan is conferred eventually on the entire human population, although I do not think this alone, without significant cognitive and emotional enhancements, will result in dramatic improvement of the human condition. Indeed, the limited distribution of radically extended lifespan might engender a reaction sufficiently severe as to eliminate the majority of extendeds, and then proceed to eliminate the practice and practitioners of virtually all natural science as enemies of humanity. This dismal scenario might not signal the end of humanity, but it would minimally herald a new dark ages.

In recent times there have been various worldwide bans of varying success (e.g., DDT, above-ground nuclear weapons testing). We might hope that a successful ban on anti-aging technology can be initiated, at least until research into improvement of human cognitive abilities and emotional characteristics bears fruit. Perhaps these provisos are best coupled with improved educational levels and overall economic parity among nations. I have doubts as to whether such bans can be enforced, but I must in good conscience deplore the libertarian viewpoint that those with the means to purchase radically extended lifespan ought to be allowed to do so, at their own personal risk. As with tobacco smoking and some other ostensibly personal choices, one's actions have effects on others. A world, a society, a neighborhood, even a family, of extendeds and nonextendeds is to my thinking a dangerously unstable situation. Of course libertarians will respond that my stance exemplifies misplaced confidence in the ability of government bureaucrats to manage a complex situation. The unsettled nature of this debate marks what I believe to be the upcoming storm, should likely events transpire.

Suppose that if there is an "upcoming storm," it abates. Assume that radically extended lifespan becomes available; that it is not universally adopted by humankind, but is available worldwide; that those possessing it (the extendeds) are in the main politically in control of their governments and societies; that whatever friction exists between extendeds and nonextendeds has settled to mutually tolerable levels. I view this assumed settled state of affairs as a comparatively happy one, because I favor our evolution to beings with radically extended lifespans. Yet as I indicated above, I do not think this evolution will be untumultuous. Further, I believe that if circumstances are such that nonextendeds outnumber extendeds and are in political power, and cannot use that power to become extendeds (imagine whatever limited-supply scenario fits this), then extendeds will be imperilled and likely persecuted.

The question at this point, granting the assumptions made at the beginning of the preceding paragraph, becomes one of treatment of the nonextendeds by the extendeds. I offer the comparison that this contemplated situation resembles, in significant respects, the treatment of indigenous people by those who have control of the lands formerly inhabited solely or mostly by the indigenous. Similar socio-political issues are raised in both situations. These include the apportionment of natural and developed wealth; the access of the indigenous to political institutions; the degree of autonomy for the indigenous. I do not believe that these issues have been resolved satisfactorily in the cases with which I have some familiarity: those of the Western Hemisphere. The United Nations Declaration on the Rights of Indigenous Peoples, adopted in 2007, evidences the over-compensatory spirit that signals the ineffective, albeit well-intentioned, attitude of many towards the problem (it is noteworthy that the four countries voting against adoption of the Declaration—Canada, Australia, New Zealand, and the United States—have had historically large confrontations of "settlers" and indigenous). Article 11–1 of the Declaration begins:

> Indigenous peoples have the right to practice and revitalize their cultural traditions and customs.

No restrictions are placed on these practices. Article 24–1 begins:

> Indigenous peoples have the right to their traditional medicines and to maintain their health practices . . .

We can imagine scenarios in which a virulent disease begins among an indigenous population. Imagine that means exist for reversing the disease, but such means are not available to "traditional medicines," and the disease threatens to spread to the entire population, indigenous and non-indigenous, and become a disabling epidemic. This is not a far-fetched situation. Surely, the non-indigenous have some claim to intrusion in violation of the above Articles.

Just as surely, this Declaration is rooted in outstanding injustices and continuing mistreatment of the indigenous. I have the hope that in the somewhat analogous, speculative situation in which the extendeds take the place of the settlers, the nonextendeds of the indigenous, matters will be better. This hope is based on my beliefs regarding the recognition of progress. While I cannot detail and defend these beliefs at length, nor the core notion of progress, I will attempt to indicate my position as follows.

I believe that on balance and all things considered, the history of human lives, as mediated by scientific and technological developments, has been in terms of value an upward progression. We do not live as we once did, say, twenty thousand years ago. If the counterfactual situation in which we communicate with our distant ancestors were miraculously to occur, it is less than obvious that we could persuade our ancestors of twenty thousand years ago to adopt our current mode of living. Yet given time, I believe these ancestors would opportunistically adopt aspects of our technology that they could operate, and little by little they would be assimilated into our society and culture in full, recognizing the overall superiority of it. I believe that had American history occurred differently, had greed and mistrust been less, and had intuitively fairer dealing on the part of European settlers been greater, a fuller assimilation of the Native American population would have occurred. I am aware that these statements will provoke anger in some, and a legitimate call for their defense on my part by many. Unfortunately, it would take another book to accomplish the requisite defense, if it can be accomplished.

The direction I'm headed should be clear. I am supposing that the population of nonextendeds, living contemporaneously with the politically stronger population of extendeds, will be assimilated into the latter group. The presupposition underlying this is that intuitively fair dealing will exist between these populations. Perhaps they will have learned from history, although to my mind there is little evidence that such learning regarding analogous matters has occurred. Rather, I envision the situation thus. Given the supposed circumstances, the nonextendeds are likely to be those whose cultural traditions and/or socio-economic circumstances have kept them at some remove from the immersion in what I loosely term "the scientific and technologically mediated lifestyle." The extendeds are likely to include in their midst a sizeable number of those whose cultural and educational influences incline them towards tolerance of and, to some extent, respect for the manners of the nonextendeds. I have stated above my belief that, were the nonextendeds in control, the outlook for the extendeds would be worrisome. Given the assumed contrary situation, I am hopeful of what I consider a positive outcome.

That matters will unfold to the point of the extendeds in control of a worldwide situation of extendeds and nonextendeds is highly speculative and problematic. I have indicated other possibilities in the course of this chapter. My fear is that the means of radical extension of lifespan will be discovered before humankind is prepared to implement properly the discovery. If the manner of radically extending human lifespan is not managed in the most rational fashion there will be conflict. We can then only desire that as the resulting turmoil subsides, what has been humanity will come to an understanding of the further requisites, such as cognitive and emotional improvement, and strive to discover and implement these in a complementary manner. When this is done we will be the better for it, but we will no longer be human.

NOTES

1. A sampling of such research is found in Stanley Shostak, *Becoming Immortal: Combining Cloning and Stem-Cell Therapy* (Albany: State University of New York Press, 2002).
2. In Lawrence Krauss, *The Physics of Star Trek* (New York: HarperCollins, 1995).
3. A sampling of competing theories is found in Marie-Françoise Schulz-Aellen, *Aging and Human Longevity* (Boston: Birkhäuser, 1997).
4. Asher Seidel, *Inhuman Thoughts: Philosophical Explorations of Posthumanity* (Lanham, Md.: Lexington Books, 2008), 31.
5. J. S. Mill, *Utilitarianism* (New York: Bobbs-Merrill, 1957), 25.

Chapter 4

The Colors of Life

Cleanthes:

. . . Health is more common than sickness: Pleasure than pain: Happiness than misery. And for one vexation which we meet with, we attain, upon computation, a hundred enjoyments.

Philo:

But allowing you what never will be believed . . . that human happiness in this life exceeds its misery . . .

(Hume: *Dialogues*)

In this chapter I want to consider an important challenge to the suggestion that humanity transition to posthumanity. In brief, the challenge is to show that having transitioned, we (or our posthuman descendants) have a desirable existence. Given that I consider such existence to be of indefinitely extended duration, the focus question is the manner in which this existence can be continually purposeful, significant, and enjoyable.

First, some remarks about the present.

There is a standard course in the Introduction to Philosophy repertoire. Not all departments offer it, and not all departments that offer it title it as follows: Philosophy of Human Nature. The content admits of various approaches. One approach: people are naturally good, naturally bad, primarily social, primarily individual, primarily dependent upon their environment for their character, primarily dependent upon their genetic endowment for their character, primarily dependent on their freedom from natural determination for their definition. Another: people are primarily material, primarily immaterial, or a combination of both. Yet another: people are best understood as conditioned organisms, best understood as biological robots under the control of a computational system, best understood as an uneasy balance of appetitive, emotional, and cognitive characteristics.

And there are yet other approaches that comfortably fit both the topic and our varied intuitions. What human beings are is a debatable, complicated matter at whatever scientific, metaphysical, or commonplace level the question is pursued. There is an advantage to pursuing the question at either the natural-scientific or metaphysical level; the answers given at those levels are more general, in the sense that they admit of less exceptions. Towards the commonplace, answers readily admit of exceptions. People are social, but what of lifelong hermits? People are desirous, but what of lifelong ascetics? And so forth. Yet whatever the shortfall from universal, or near-universal, applicability attending commonsense-level approaches, natural-scientific theories of personhood fail to satisfy the craving for the sort of answers that are supplied at the more commonsense level. Perhaps to a lesser extent metaphysical theories of personhood also have such lack. This is so, however faulty such commonsense answers are shown to be on critical inspection. The proper conclusion might be that we need get over our desire for commonsense descriptions/explanations of personhood. Or perhaps that we need redouble our efforts to find a satisfying, widely applicable theory at that level.

I have indicated in a previous chapter that I do not believe we possess several of the deeper secrets of what we are. I have also indicated that I believe such knowledge, if possible, will be of natural-scientific form, although not necessarily covered in full by the extant natural sciences of physics, chemistry, and biology. In that same chapter, "The Future of Ethics," I maintained that we do not yet have a satisfactory ethical theory as to the best life to lead. As I said there, this lack is at least partially due to the extent of our ignorance of our own species-nature. Most of us think that some lives are better than others, and that some moments in a given life are better than others in that same life. However, with the exception of grotesque comparisons, it is difficult to generalize these judgments to the degree that all, or nearly all, thoughtful people will concur. Such difficulty is an indication of our lack of a satisfactory ethical theory, if success is measured by anything like the standards to which natural-scientific theories are held.

Given these observations, the supporters of a posthuman existence might attempt an easy evasion of the criticism that it is difficult, if not impossible, to show that posthuman life is desirable. Their obvious argument would premise the difficulties, if not the impossibility, of elucidating the optimal human life, and conclude that one should expect similar difficulties in specifying the best posthuman life. Further, given that what is ultimately at issue is the desirability of posthuman life, competing claims regarding the ends of human life do not provide a stable logical platform from which to launch universally accepted speculations as to the general desirability of a posthuman existence, let alone specified forms of such existence. The strategy indicated here is to note that a major reason for not transitioning to posthuman life is the supposed inability to indicate a desirable form, or forms, of posthuman life, which inability does not differ significantly from that of specifying such a form, or forms, for current and foreseeably future human life. But if the difficulty is common to human life, as currently understood, as well as contemplated posthuman life, then this difficulty should not by itself count against the latter.

Unfortunately for the supporter of posthuman life, the critic can grant this argument because the critic relies on a more positive observation; that posthuman life is generally regarded as undesirable by most humans. This is not simply to espouse the negative proposition that specifying the common desirability of posthuman life is lacking. Rather, it is to argue that any portrait or sketch of posthuman life, if given honestly and with some detail by proponents of such life, will be rejected by the overwhelming majority of humanity.

In putting such an argument in the mouth of the critic of posthumanity, I am relying on the reader's familiarity with posthuman dystopias such as occur in *Brave New World* and *The Time Machine*, together with various science-fiction scenarios of post-apocalyptic deformities. But not solely such vivid and easily countered examples. There are subtler arguments resting on premises such as the not fully accountable, variegated, perhaps family-resemblance qualified whatness of being human. There have been notable examples of tampering with a naturally occurring species or phenomena in the interest of improvement, with unforeseen and deleterious results. Our humanity seems far too precious to risk similar consequences.

Humans typically want to remain humans. Confronted with ideas of radical change, humans often fix on what they will lose, or what they possess that will apparently be altered so as to no longer be familiar, and reject such alteration. It requires either the perception of dire circumstances, or the promise of immediate-to-near-term profound and certain enhancement, to initiate acceptance of radical change in the majority of humanity. To the extent that neither of these circumstances is present, arguments that proponents of posthumanism consider persuasive fall on deaf ears.

To this point I hope I have been fair in representing the opposition to the suggestion that humanity transition to posthumanity. There may be powerful arguments in support of such opposition that I have not indicated. I have not, for example, considered theological objections to posthumanism. While I categorically reject any such arguments as worthy of consideration, I am aware that others do not. I am relying on the "unforeseen consequences" argument, supported in part by my admission that there are significant aspects of our species-nature we currently do not know, to cover much of this opposition.[1] I am hoping that the remainder is largely, if not completely, addressed through a consideration of the argument that posthuman life is not worth transitioning to, because it lacks the diverse joys of the human life from which it departs. It is this latter objection I will confront in this chapter. Of course, it is impossible to address fully the specific unforeseen consequences of transitioning to posthumanity for the logical reason that, if addressed, such consequences are not unforeseen. Granting the unforeseen consequences of transitioning its proper respect, it should be noted that there are likewise unforeseen consequences of remaining human. For the continuation of those qualities we value most, assuming a rough unanimity of valuation is possible here, the time may come when as a species we must either transform ourselves or face unfortunate hitherto unforeseen consequences.

At this time any account of posthuman life is speculative. An additional difficulty regards the specification of criteria that allow use of the term "posthuman," minimally understood to mean "other than human, yet in some manner evolved from human." The latter part of this minimal understanding of "posthuman," while less problematic than the former half, is not without issues. Even if restricted to biological evolution, questions arise. For example, suppose biological humans create a biological entity by some sort of laboratory procedure dispensing with human DNA. Suppose further that this entity is endowed with human-like cognition, perhaps via an uploading of human neural connectivity patterns onto its nascent brain or brain-substitute. In what sense is this entity an evolvement from its human creators? Whatever difficulties attend this question are multiplied by imaginative consideration of non-biological "progeny."

Much popular science fiction suffers scant difficulty presenting currently nonexistent technology. Bodies "teleport," the speed of light is transcended, various antigravity devices allow remarkable freedom of movement, and artificial gravity is sometimes provided by means not currently understood. What is striking in all this is the presence of humans, indeed mortal humans, with biological, psychological, and sociological characteristics indistinguishable from one's present-day peers. Often enough, these future humans are in proximity to nonhuman creatures having comparable psychological and sociological characteristics, although storylines are usually best served by amplifying some of such characteristics among the nonhumans, and suppressing others.

This imaginative lack gives evidence of the difficulty attending speculative thoughts as to the possible nature of posthumanity. Simultaneous evidence is given supporting the criticism that posthumanist thought is not the manner in which humans want to consider their present or future selves. Still, the part of this evidence not due to imaginative laziness might be ascribed to psychological inertia. One could argue that advertising techniques allow for the imagining of a posthuman future that might appear attractive to many, however superficial the presentation. Entertain for the moment an artfully contrived message to the seventeen- to twenty-five-year-old segment of the male population of the developed sector regarding the enhanced sexual experiences of some supposed posthuman form. Or consider that in ordinary circumstances few people are willing to risk violent death on a daily basis, yet many willingly undergo such risks in wartime for questionable rationales.

Granting the force of at least some of the arguments mentioned in the preceding paragraphs, the challenge remains to present a positive characterization of posthuman life. I want to take up this challenge, but first I want to stipulate some terms. These terms are occasioned by considerations presented in two preceding chapters. There are the metaphysical and epistemological constraints emphasized in a preceding chapter: consciousness, externality, and continuity. And there is the general constraint presented in "The Future of Ethics": that "pointers," as these are understood from the human perspective, serve to guide the initial transhuman phase.

Thus far it appears my stipulations are given to preserve as much as possible of human-like life, thereby easing my assigned task of showing the joys and diversities of posthumanity. The following should serve to mitigate that appearance.

Near-term posthumans are understood to have indefinitely extended lifespans. With this stipulation I indicate two noteworthy circumstances. First, that the actualized possibility of such lifespan is likely an initial transhumanist condition. Put differently, within the next few decades humans may have the option of radically increasing their life expectancy. This minimally assumes continued progress in the biological sciences, which in turn presupposes a sufficiently stable socio-economic environment for the sorts of research having such results. Such assumptions are not, of course, vouchsafed. Second, that entities living indefinitely extended lives—ourselves or our descendants—will at some point cease to be human.

It will not serve any useful purpose to engage in terminological disputes regarding the term "human." There are clear cases of the human and the nonhuman, and there are problematic cases of each. I believe that any biological descendant of ourselves that lives an indefinitely extended life will be at sufficient biological and psychological variance, if not also sociological variance, with present-day human beings as to warrant the general term "posthuman," with the understanding that what is posthuman is to some extent not human. If one is enamored of the term "human" to the extent that one deems human an electro-mechanical device that passes the Turing test and in various ways successfully copes with its environment, I will not insist on "posthuman" entailing "nonhuman." I only request permission to use the term "posthuman" as signaling some significant distinction between entities that have had significant commonalities for the past roughly one hundred thousand years, and the entities that I am considering.

Consider a biological entity, either ourselves or what our descendants have become, leading an indefinitely extended life. There are various general possibilities. Perhaps the possibility requiring the least difference to present humans is the following. Imagine that present day humans can, by pharmacological means, retard and reverse in a controlled manner the hitherto effects of biological aging. Other than taking the appropriate daily (or weekly or monthly, etc.) oral medication, these entities look like us, cognize like us, experience like us, and have appetites similar to our own.

I think that some people regard the possibility of indefinitely extended life in this manner. If so, their regard is arguably ill-conceived. Much of our biological apparatus, however rejuvenated, cannot withstand extreme impacts whose probability increases with longevity. New diseases, genetic malfunctions, and environmental misfortunes such as famines, argue for a withdrawal of the minimal-difference assumption. Stronger assumptions involve various implants, prostheses, uploading of brain contents, possible temporary cryogenic storage, and so forth, perhaps to the point of a major or total relinquishing of biological constitution. Of course, one need not accept all, or even the major part, of these stronger assumptions. One might choose to consider the possibility of indefinitely extended life, other things being equal. That is, one might allow for various possibilities of termination of an indefinitely extended lifespan, the duration of which continues only if those possibilities are not realized. Such a move allows for more tolerance regarding application of the term "human" to such entities. Apart from the linguistic result, this move runs counter to my undertaking of showing that posthumans, as we are likely to consider them, have joyful and diverse lives, in that it allows too much reference to the joys and diversities of current human life. Still, this move will receive some further attention below, and in the following chapter.

Some may reply that all the stronger assumptions of the previous paragraph granted, the entities under consideration could yet be psychologically constituted much as ourselves. Given the sorts of modifications mentioned in the preceding paragraph, this reply seems doubtful. Even waiving such modifications, there is reason to be skeptical of the psychological similarities between ourselves and these speculated descendants. Consider a deep and pervasive fact about adult human life, our sexual drive.

For various reasons, I will not discuss such matters as the differences, if any, between male and female sexuality; same gender preferences; sexual abstinence; and sexual perversion (if such a category be deemed admissible). The ways of sex are many. The ramifications of our sexual natures as they generally involve other aspects of our lives are often apparent, controversies regarding details of such involvement granted. Various humans, for various reasons, have apparently been able to forswear at least some aspects of their sexuality, aspects which the overwhelming majority of humanity has not renounced. But to be fully nonsexual at all levels, conscious and unconscious, psychologically as well as biologically, seems beyond possibility for virtually all humanity.

A cognitive psychologist friend of mine puts it simply: "We're hardwired for sex." I would add that we are in a number of ways culturalized to specific aspects of sexual desire and practice, and that our propensity to such culturalization, as with many other matters of social influencing, is "hardwired." We engage in sexual practice partially by virtue of our biological instinct to procreate, an instinct whose evolutionary survival value is a near-tautology. We also engage in sexual practice for reasons of pleasure, regardless of any conscious desire to procreate. Human life, fully absent of sex in all its manifestations—biological, psychological, and sociological—would be sufficiently different from what it is and has been so as to raise the following two questions: would such life be appropriately termed "human," and would it be worth living? I have indicated my relative indifference to the first question. I will be addressing the second question below.

If we evolve to entities that have lives of indefinitely extended duration, there is a question to confront that is not always confronted in the literature. To what extent would these entities—who by some of the above speculations would not differ significantly from us, even given their indefinitely extended lifetime—desire offspring, what we currently term "children?" Perhaps the answer is obvious; if they do not differ greatly from us, they would desire children as much as we desire them. There is an additional consideration, however. In living indefinitely extended lives, our planet would at some not-too-distant time lack the space in which to support the ever-increasing population. Assuming various further possibilities, such as colonization of other planets, shrinking of bodily size, and, as referenced above, various calamities that eliminate entities not sufficiently enhanced to sustain what were termed above "extreme impacts" and "environmental misfortunes," we can imagine an enlargement of the room currently available for children. In the longer run, however, on the assumption that children increase the number of entities with indefinitely extended lifespans, it appears unavoidable that the human experience of having children, as now understood, will need be abandoned.

Humans engage in sexual activity for pleasure, as well as for procreation. Even if children are no longer a possibility, it seems plausible that humans will continue to enjoy sexual activity. Yet I submit that removed from the possibility of procreation, having bodies resistant to environmental shocks, it is possible that posthumans might in time fully lose their sexuality and become, not unisexed, but simply nonsexed.

I want to explore the claim that nonsexed posthumans could not have lives that we humans would consider joyful, diverse, and in short, worth living. I do not believe this claim is correct, but it must be acknowledged that any such discussion is speculative. Of course, this speculative aspect cuts on both sides of the issue. I am willing to grant that almost any entity significantly resembling current humans in its biological and psychological characteristics would find a sexless life of extended duration lacking; perhaps so lacking as to be not worth living. What is supposed here, perhaps contrary to laws of nature, is an entity biologically and psychologically virtually "twin-earth" similar to humans, but having an indefinitely extended life at, say, the biological equivalent of a thirty-years of age human, and wholly lacking in sexuality. It may be that such supposed entities have sufficient variance so that some of them do find such a life worth living. For the little that can be confidently concluded from my sketchy assumptions, it may be that all or at least most such entities find their extended lives joyful. But I think it unlikely that most contemporary humans would consider such life worth living.

We have arrived at an impasse. Many present day humans desire a physically healthy and comparatively youthful mortal life, indefinitely extended. That is, many humans complain of aging and its attendant ills, and at almost any given moment do not desire death. Yet many of these same humans do not desire childlessness, and scarcely any desire sexlessness. In a previous work I suggested that this impasse would likely be overcome by our choice of the alternative of extended life, as it became available to us.[2] Implicit in this suggestion is the judgment that most humans either would not be mindful of the relatively distant consequences of this choice, or that their present desire for extended lifespan would outweigh thoughts of later consequences. In respecting the "pointers to the best choice" response given in a preceding chapter to the question "whither humanity," I shall attempt to improve on my earlier suggestion.

Since I do not suppose that we or our descendants will be living lives of extended duration with our present biological and psychological makeup, I want to add some assumptions about posthuman mindfulness. Perhaps most significantly, such entities will have cognitive abilities in excess of our own. Rather than attempt a detailed sketch of these abilities, let the following formula be adopted. For any current human cognitive feat or ability subject to measurement, increase the average ability of posthumans to that of superior human performance by threefold. This is admittedly vague and crude, but it lends some intuitive purchase on the assumption of cognitive superiority of the posthuman. Further, all else held constant, I believe many current humans on first thought would consider such transformation desirable.

Second thoughts are another matter. Holding all else constant includes holding emotion-driven attitudes towards others constant. What of anger, greed, jealousy joined to the assumed posthumanly powerful cognitive abilities? This is a troublesome mixture. What of those calculating, clever (in the Aristotelean sense, *Ethics* 1144a 20-30) individuals now possessed of enhanced cognitive powers? One might hope that widespread enhanced cognition would swiftly lead to a world in which posthumans overcome various individually and socially harmful states of mind. Allow me to sidestep the difficulty of elaborating which are such states of mind. Indeed, as I stated in "The Future of Ethics," I doubt that we are currently in possession of a satisfactory psychology and sociology which would adequately support every ethical judgment of harmfulness. This is not to say that we lack strong beliefs and intuitions regarding such matters. Rather than say any more regarding this, I refer the reader back to the relevant discussion in that chapter.

I share the hope that the above assumed enhanced cognitive powers lead to more ethical individuals and a more just society. The initial step in such improvement might be discoveries regarding what are now considered to be various psychological aspects of persons. Among other benefits, such discoveries could indicate the means for those psychological transformations which bear upon social improvements. What I am supposing here is an upward spiral of positive events in which, as humans and their posthuman descendants learn more about their own psychological processes, they learn more about the social effects of various implemented changes in those processes, and they learn more about the manners in which those social effects have psychological significance, and so forth. In particular, the increasing cognitive powers of these entities facilitate the discovery and implementation of progressive social schemes.

Humans have borne a self-concern that has gone beyond its evolutionary origins of self-preservation, and self-enhancement for mating purposes. Philosophical social thinkers have variously addressed this self-concern, some generally supporting it, others seeking to mitigate it. In his "second wave" discussion within Book V of the *Republic*, Plato's Socrates speaks of a social class of guardians transformed by education so as to be largely group-regarding in their desires and activities. In *Utilitarianism*, Mill makes a similar point intended for all members of society:

> In an improving state of the human mind, the influences are constantly on the increase, which tend to generate in each individual a feeling of unity with all the

> rest; which, if perfect, would make him never think of, or desire, any beneficial condition for himself, in the benefits of which they are not included. If we now suppose this feeling of unity to be taught as a religion, and the whole force of education, of institutions, and of opinion, directed, as it once was in the case of religion, to make every person grow up from infancy surrounded on all sides both by the profession and the practice of it . . .[3]

The hope for such widespread state of minds does not end with Mill. In more recent times the attempted implementation of it has not always yielded positive results, and in no case has it been successfully implemented among a considerable population for a century-long duration. The failure of this vision to date cannot wholly be ascribed to humans' apparently contrary psychological nature. It may be that any economically viable human society demands organizational structures, and these structures demand various authoritarian relationships which counter attempts to instill the spirit of unity among all members of a society. Mill himself worked for the East India Company, whose operations did not reflect attitudes of social unity inclusive of the opium-addicted Chinese population, whose cravings for opium were nurtured and catered to by the Company.

Still, one might harbor the hope that were certain changes effected in our psychological natures, our descendants might come to realize a society of entities with both enhanced cognitive powers and superior moral characteristics, however vague and intuitive the latter are in our current understanding. Further, the currently often antagonistic social relations of production might be eased in a number of ways. Various laborious activities might be done by increasingly capable machinery, for example.

Old habits of thought need to be overcome, if a vision of unprecedented, species-wide social unity is to be achieved. Of course, large numbers of people do not favor this vision. Those who believe in a theologically supported system of castes, to name one such group of people, will resist the leveling in social status such an ideal supports. Those who believe in other forms of privilege by birth will have similar disapproval. These two examples of opposition are hopefully modes of thinking that are largely discredited in the eyes of most of humanity. However, there remain other pervasive manners of belief which do not allow for schemes of social unity of the sort envisaged here. Those who believe in the value of competition, in "the virtue of selfishness," will raise objections. It is often claimed that various people have innately differing capacities, and these differences should be reflected to some extent in different socio-economic roles. In the *Republic* Plato famously presents and defends this viewpoint, arguing for a social cohesion that rests on members of the different groups recognizing and accepting their different roles. We should note that Plato himself doubts the long-lasting viability of his proposed ideal society. Also noteworthy is the founding of this ostensibly ideal society on a division of labor necessitated by then extant modes of production. Given the relatively scant surplus to support a ruling class of the wise, it is intellectually convenient for Plato to allow the commonplace assumption that different people are innately suited to different modes of life. One wonders in what manner Plato might modify his vision of an ideal society, were he aware of the greatly expanded surpluses capable of being generated by modern modes of production. Regarding the competitive spirit and its claimed benefits for those less competitive, one can imagine a society in which the members are in unified competition not against each other, but rather against ignorance, disease (or deterioration), stagnation, and other deficits that will continually confront them, however scientifically and technologically advanced they become.

What has been said will almost certainly be deemed unsatisfactory by those whose ideals conflict with the egalitarian strain of thought manifested in the preceding paragraph. There are those who would object that a childless and sexless future, whatever else this future promises, is unacceptable. For all others, however, there is the lingering question: what sort of life, or lives, can compensate our descendants for a childless and sexless life, let alone surpass the lives current and foreseeable humans live, or could live under the best of human circumstances.

By "near-term posthuman" I intend those entities comparatively close to the transition from the human. It is understood that there might never be such entities, or that they might be post-nuclear apocalyptic horrors, or any number of less-than-desirables. I shall use the term "posthuman" in the ensuing discussion to refer to these near-term posthumans. I shall further assume that these posthumans have been fortunate to the extent that they lead lives many present-day humans would judge desirable on careful reflection. It is my task to make good on this assumption.

In discussing the desirability of posthuman life, I shall examine three major components of such life: the cognitive, the ethical, and the aesthetic. I do not assume these three components exhaust the content of speculated posthuman life, nor that they are the sole major components of such life. Moreover, I take this choice of categories to be to some extent arbitrary, in that other cross-categorizations are possible. Further, I do not suggest that these categories are exclusive of one another. There exists, rather, much interrelation. The choice of a category such as the cognitive likely suggests to some a traditional pairing with the affective and the appetitive. Aspects of this latter pair of categories will be considered as manifestations within the three major categories chosen above.

Philosophical debates regarding the best sort of human life are unending. Given the notion of "pointers" from the human to the posthuman, indications are that these debates carry over to considerations of the best posthuman form of life. These indications are not logically necessary, in that some sympathetic both to posthumanism, and to the notion of pointers from the human to the posthuman, might consistently expect a reordering of priorities in the transition. Such expectation has already been supported by the proposal of posthuman childlessness and sexlessness. With awareness of the unsettled nature of the debate, I want to make a case for the cognitive life; not exclusive of other activity, but as the most noteworthy aspect of present life and of the near-term posthuman life I contemplate.

By any understanding of the term, the "cognitive" is a large and diverse category. I shall be brief in detailing my usage. In considering the term as it applies to humans and envisaged posthumans, I make a loose division of cognition into the theoretical and the practical. While I find this terminology convenient, it is misleading as I am employing it. I do not intend the standard philosophical direction of ethical concern in my choice of the term "practical." Rather, I mean the everyday, the sort of cognitive activity that allows for our routine being-in-the-world as embodied selves, within a socio-historical and cultural framework. This, of course, covers a lot of ill-bounded, incompletely understood ground.

Allow me to appeal to common intuitions regarding the domain of the theoretical. I assume without demonstration that the boundary between what I am terming the "theoretical" and the "practical" is indistinct, although there are clear cases falling on one side or the other.

In making my case for prioritizing the cognitive life, I propose the following exercise: consider what you would least want to lose. I believe without any evidence other than my intuitions that most people would answer, on a moment's reflection, that they would least want to lose their life, especially those that do not have strong anticipation of a favorable existence upon their biological death. Let the question be modified to read: what would you least like to lose while retaining biological life? I suppose a range of answers to this question. Some will preference various property items, including financial wealth. Some will respond generally with terms such as "health," "love," or even "anonymity." Some will speak of body parts. Accepting this varied list of prima facie responses, I maintain that careful, and admittedly directed, questioning will lead many people to aver that what they would least want to lose is their cognitive mind. Put another way, many people will on reflection consider an affliction such as Alzheimer's disease to produce the greatest loss a person can suffer.

I would answer the question in that way; I would least want to lose my cognitive mind. I would add that in losing my cognitive mind I have in effect lost my life, because my cognitive mind is a large part of what I term "myself." Further, I am assuming that I have not lost myself to become another human self, as in imaginary science fiction scenarios. That sort of case would require more investigation. Rather, I am considering the case of losing an irreplaceable cognitive self, and coming to the judgment that this is the worst loss imaginable.

I suppose many humans, if not most, given to theoretical cognitive activity would agree with me on this point. Those less inclined to theoretical cognitive activity, including some of my students whom I have queried on this, are initially inclined to responses that indicate fear of blindness, castration, amputation of limbs, quadriplegia, and other physical afflictions. When steered to the response I favor, I have at times encountered the following objection. Granted that losing one's own self is a terrible loss—perhaps the greatest loss—loss of cognitive powers is not fundamental to loss of self. Rather, loss of self entails loss of memories that constitute one's life history. That loss is what makes conditions such as Alzheimer's terrible.

Couched in philosophical terms, this objection considers loss of self as loss of "center of narrative gravity," perhaps even loss of "transcendental ego." I agree that these losses constitute loss of self, but I do not agree that these losses are other than the loss of which I speak. Such losses are for the most part cognitive losses. That the sense of self ordinarily possessed by humans is cognitively charged is a proposition requiring little argument. Descending the phylogenetic scale, in what manner does a great ape's sense of self compare to a chicken's sense of self, or to a spider's? Given our current philosophical and scientific understanding of such matters, the greater part of our answer to these questions is speculative. I speculate that as examples progress downwards on the phylogenetic scale, awareness of self becomes more a matter of reflex-behavior, increasingly momentary. I grant that most humans' sense of self is not bound to their theoretical cognitive activity, but I reject the suggestion that it is not bound to what I am naming their practical cognitive activity. Even a character as cognitively impaired as Faulkner's Benjy Compson (the idiot from *The Sound and the Fury*) has memories having organization according to linguistic categories. Absent these, and absent the sensible groupings that a nonhuman mammal might use to organize such memories as it has, it is difficult to conceive a sense of self that a human could recognize.

The logic of the above case is less than certain. It relies on the equivalence of "that which we would least want to lose" to "that which we value most." Minimally, the former would need to imply the latter. Yet these two phrases are not equivalent, nor is there a relation of necessity from what we would least want to lose to what we value most. These are intentional phrases, involving that which is held in a type of regard. There is no objective contradiction in negatively valuing something and not positively valuing its negation. It might be an objective fact of nature that some individual despises both X and despises the negation of X. Of course, such an individual might typically be considered irrational on this score, but such irrationality is then an objective fact.

Relaxing the rigorous requirement of logical necessity, the previous argument for prioritizing cognitive activity has some intuitive strength, but it will not fully serve my purpose. Its lack is due to two faults. First, it is insufficient to note, as the argument does, that much of our sense of self has a cognitive aspect. There are arguably other pervasive aspects to most humans' sense of self, such as our sexuality, which I have argued will not successfully carry over to posthuman life. Second, I believe that the most significant aspect of our posthuman life will be what I am terming "theoretical cognitive activity." I shall attempt to defend this belief.

To elaborate, it is noted that posthuman life demands practical cognitive activity. In the sense of "significant" that is captured by "necessary," practical cognitive activity is arguably significant for each individual posthuman, whereas theoretical cognitive activity is not. In deeming theoretical cognitive activity the most significant aspect of posthuman life, I am using "significant" in the sense of "desirable," and by considering theoretical cognitive activity as being most significant, I mean that it is the activity that will be desired above all other activities. "Being desired above all else" is a phrase which itself is in need of elaboration, but I shall stop here and leave the remainder to intuition.

It should be recalled that the case for transitioning to a specific posthuman form rests on what I have termed "pointers" from our present circumstances. That is, it should be shown that, given that we have reason to transition, we want to transition to entities having characteristics X, Y, Z, rather than characteristics which negate X, Y, Z. If wanting certain characteristics is in conflict with wanting other characteristics, then we ought to consider which of the competing characteristics we want more. So it is with indefinitely extended life, which I believe is in conflict with procreation and hence ultimately with sexuality itself.

Can it be maintained that most presently existing adult humans value theoretical cognitive activity highly, perhaps above all else? In the sense of the question that intends "value it for themselves," where the criteria of such valuation include daily behavior, most would aver that the answer to this question is negative. Allow a modified question: do most such humans highly value this activity for their children? The answer here, I think, is less decidable. Many parents of children will weigh their answer, taking into consideration their expectations for their existing children, who in many cases have already indicated their own life-choices. In fairness to the point of this question, let us consider a final version of the question: would most humans choose such a life for their yet unborn children, if their present socio-economic circumstances were not of concern in making the choice?

No one knows the answer to this last question. The answer remains unknown even if the audience for the question is filtered so as to exclude those humans who have scant exposure to theoretical activity, or those for whom theoretical activity is constrained by theological dogma. Many parents will likely respond that the happiness or peace of mind of their children is their primary wish. I submit that this latter response is ill-considered. An individual with severe cognitive impairment who is nevertheless tranquil and in some sense happy is not what most parents desire for their children. Admittedly, freedom from agonizing pain or severe psychological duress is an unproblematic wish of parents for their children. That wish granted, what sort of life is most desired for the yet-unborn? At the level of generality occupied by choices such as the life of mind there are other choices. The life of beauty—the aesthetic life—is an alternative choice, as is the life of power, of wealth, of respect, of recognition. Of course, some of these alternatives have plausible intersections with each other and with the life of theoretical cognitive activity. Further, some of these alternatives may be diminished in parents' eyes when ethical considerations are raised.

I suppose without much argument that many parents would be most desirous of their children being recognized as superior thinkers. I suppose that their pride as parents would be at its peak were their child a Pasteur, a von Neumann, or an Einstein. I admit that their pride might be equally great were their child a Beethoven or a Rembrandt. Waiving for the while the issue of the boundary between artistic genius and genius of theoretical thought, there are unnecessary interferences in my suppositions. Success and recognition attend all the figures I have mentioned, and prideful parents typically wish these attend their children. If the parents of my hypotheses are asked to compare the legion of artists who labor unrecognized with the similarly large group of those engaged in theoretical thought—the scientists, mathematicians, humanists, and others—the choice of the majority might well be for the latter. Perhaps this will be their choice even if prevailing economic circumstances (artists starve, scientists don't) are somehow taken out of consideration.

Thus far, my strategy of supporting the proposition that theoretical cognitive activity will be the most highly valued aspect of posthuman life, by showing that such is the life most human parents wish for their children, has fallen short. There have been too many questionable assumptions, some explicit some implicit, and too little exploration of details. More speculative details of posthuman theoretical cognitive activity will be given below. But I want to try buttressing my "parental choice" argument further, because this is not merely an argument from analogy (as parents desire for their children, so we desire for our posthuman descendants—assuming we desire posthuman descendants at all, rather than continually human descendants). If reasons and/or circumstances lead us to desire a posthuman future for our descendants, we are deciding such for our children, or our grandchildren, or our great-grandchildren, or whenever further the transition occurs.

In vernacular terms, parents typically want their children to be good people. They want them to be successful people, and they want them to be happy people. The parental wish for happiness has been addressed above. Regarding moral goodness and success, it may be a departure from the commonplace if phrased as follows, but can it be said that most parents want their children to live in a society in which a paradigmatically transformative yet peaceful positive social revolution has taken place? Some further elaboration of this high-sounding phrase was provided in the chapter "The Future of Ethics." With reference to that chapter, let us emphasize the notions of peacefulness and equality considered therein. Given historical and current circumstances, the wish for success of one's descendants is in part an acknowledgment of human life lived in trying conditions. At least some of these conditions are due to various psychological attitudes of humans that, in appropriate circumstances, have outlived their evolutionary usefulness. With extended life and enhanced cognitive abilities, our descendants have greater possibility of realizing a genuinely moral life as part of the well-being of all members of a transformed society. In such a society, success might not be measured in terms of competition among the society's members.

With regard to what has been discussed above, what our future will be is beyond probabilistic assessment. This epistemological reality noted, there is some illustrative value in exploring the consequences of the idealized assumptions that have been made, and are about to be made. To the point: what might an indefinitely extended life emphasizing theoretical cognitive activity be like?

Currently, the majority of literate humanity in the developed sector do not much engage in theoretical cognitive activity. Many pass their adult lives having their only contact with such activity in the indirect form of reading or listening to reports of such activity; popularizations such as are reported in the various media. These people have mostly been exposed to this thought in their educational curricula, but having passed through their interval of formal education they have not returned to it.

Allow that matters are different for our posthuman descendants. Psychological matters have been *engineered* to be different. These entities want to engage in theoretical thought, much as Einstein, von Neumann, Wittgenstein, Sartre, T. S. Eliot, and a host of others wanted and want, in various ways, to engage in such thought. Suppose that we, in the extended sense "parents" of these posthumans, want them to have such wants.

I'll begin my speculative answer to the question as to what such a state of affairs would be like by considering the current domain of natural science. Included in this domain are physics, chemistry, biology, combinations of these (e.g., bio-chemistry), and subsidiary sciences such as astronomy and geology. Also included is the mathematics applicable to these sciences. Assume that these sciences carry over to the time of our speculated posthuman descendants. Assume further that almost all these descendants minimally possess or are in the process of possessing an understanding of all these sciences at the level of a current advanced graduate student specializing in one or the other of these sciences. Assume yet further that this understanding is integrated to an extent that few, if any, current or past humans are or were capable.

This vision of posthuman engagement with natural scientific theory will seem outlandish to many. Yet among humans there have been those recognized as having abilities and interests approaching the entities in the above description. Scientifically oriented mathematicians such as Leibniz and von Neumann come readily to mind. However, I am not speculating about a society composed of Leibnizes and von Neumanns. They had human lifespans and human neurophysiology. I have been considering a society of entities who will have indefinitely extended lifespans and cognitive apparatus in excess of that of Leibniz or von Neumann. Imagine if Leibniz and von Neumann were granted indefinitely extended lifespans and cognitive enhancements (to appeal to a science-fiction vulgarity, "brain boosts"), while otherwise living within human society. Waiving likely inconsistencies and somewhat lesser difficulties in this proposal, what would be expected of these humans so enhanced? While it is impossible to predict the outcome of this counterfactual supposition, one might imagine an intellectual glory bestowed on humankind. My immodest proposal is that we imagine a future society composed of members who are all so constituted.

Often included in discussions of traditional modes of living is the aesthetic life. In the Western tradition of aesthetics notable practitioners of this life include Mozart, Beethoven, Rembrandt, and Shakespeare. The field of aesthetics is wide, as are the considerations introduced by distinctions such as high and low art, art versus craft, and performance art against plastic art. With this breadth of field in mind, I want to entertain the hypothesis that the typical posthuman of my speculation has acquaintance with all such areas, practices at least several, and is an adept theoretician (e.g., literary critic) regarding at least several of these.

Under the freedom of hypothesis modus operandi I have been using here, I want to add one more assumption. Let it be allowed that these posthuman scholars and practitioners of natural science and these posthuman scholars and practitioners of the arts are one and the same entities. While I am affording myself some guilty pleasure in so twisting the screw, let it be recalled that there is some historical precedent for this assumption; at the least, Leonardo. I am hence imagining a society of enhanced Leonardos. Or, enhanced Benjamin Franklins (recall that Franklin played and composed music).

Some readers will almost certainly view the sort of existence I have been sketching as joyful, and *joyful for them*, other things being equal. But I suppose that almost all thoughtful readers will have concerns. While I cannot address every reasonable concern, I want to try responding to some that I believe are most significant. I will then resume the thread of my speculations.

1. Indefinitely extended life appears a mixed blessing. How does one occupy oneself for hundreds or thousands of years? Surely, one gets weary of engaging in science and art.

It is not difficult to imagine one such as ourselves becoming weary of this sort of life, to the extent this life is indicated thus far. On the other hand, at any given moment most of us do not want to die. Granted, the conatus of an indefinitely extended lifespan, manifested moment by moment, seems burdensome when measured against our expectations of a human lifespan. Added to this burden is the speculated sexlessness of this extended life.

My response to this is that at least some humans, such as Einstein, Leonardo, and Franklin, might not have been bored had this lifespan and enhanced cognitive abilities been bestowed on them. To the objection that most of us are not appropriately constituted to appreciate this gift I reply that our posthuman descendants will be.

2. It appears a blank check has been issued to "engineering" whatever characteristics are needed to realize this vision of a posthuman utopia. Sober reflection on visions of such personality-engineered utopian societies, for example Skinner's *Walden Two*, at times run contrary to our initial intuitions regarding fitting ourselves into pre-conceived molds.

This important objection has two parts. One half challenges our ability to engineer whatever apparently desirable characteristic, while the other challenges both our methods and foresight in attempting such alteration.

Whether we come to possess such ability, or even to approximate such ability, is an empirical question which I shall not address here, except to note my agreement that Skinnerian conditioning is not up to that task. While I do not want to attack straw men, there is an all too prevalent concern with tampering with human nature, and in particular with anything that smacks of conditioning. It hardly needs saying that education, childrearing, and socialization all rely somewhat on what can at least loosely be termed "conditioning." We currently take mood altering pharmaceuticals and, in severe circumstances, undergo more dramatically invasive treatment such as electroshock therapy, and forms of brain surgery. These constitute much of our current repertoire of personality engineering. They are some of them pervasive, and they vary in their effectiveness.

Granted that our foresight is no match for our hindsight, and our hindsight regarding matters under discussion has its share of regrets. A moment's thought on the reliance upon treatments such as lobotomy serves as reminder here. Yet we must act in the matter of forming our next generations, whether our general approach is conservative, moderate, progressive, or radical.

3. Attention must be paid to the all else that has been held equal. These imagined posthumans will have problems with which to contend, be they mundane problems such as those of, or analogous to, home repairs, or larger issues regarding matters such as those of social, political, and economic organization.

Agreed. If I were to grant what I am unwilling to grant—namely, that solely occupying oneself with science and art for an indefinitely extended lifespan would be boring—this last concern at least partially relieves that problem. There are varieties of what might be termed "applied" or "practical" matters that will demand attention. Rather than limit further comment to a direct response to this objection, I want to return to the main discussion with which I have been engaged, and allow this return to function, in part, as a response to this last consideration.

I am supposing that much of routine work in a posthuman society will be done by machinery. Some of this machinery will be "smart" and "adaptive." Some of it might be capable of doing much, if not all, of various specialized tasks (e.g., neurosurgery) that only a select group of well-trained humans currently perform. Transferring much currently done human activity to machines raises questions associated with the possibility of broad-spectrum artificial intelligence (AI). I want to forestall such questions. They have been considered briefly in a previous chapter, and they will be further addressed in the subsequent chapter. The discussion now is under the assumption that broad-spectrum AI has not been implemented. Smart, adaptive machinery is assumed under human creation and supervision, and typically specialized in application.

These imagined posthumans will be involved in practical affairs. The quality of their involvement rests in large part upon two general assumptions. First, what are presently considered as the social and behavioral sciences will be considerably improved as to concepts, theories, and methodology. They will be well-integrated, as are presently physics, chemistry, and biology, and they will have a degree of agreement among practitioners approaching that currently enjoyed by the latter natural sciences. Along with these sub-assumptions let us add that, as with the above-speculated natural scientific activity, these improved sciences will be engaged by the entire posthuman population.

The second general assumption relies heavily on speculations introduced in the chapter titled "The Future of Ethics." Succinctly put, the assumption is that these posthumans will behave better towards one another than humans are capable of doing. This assumption cannot be made good, or even discussed in detail, until such progress is made in relevant areas of natural and social science as is broadly indicated in "The Future of Ethics." Still, we may wish to speculate provisionally regarding aspects of the character of these posthumans. Suppose, for example, that they are largely, if not fully, noncompetitive towards each other. If they are nonsexual, they will not compete in this arena. If there is no scarcity of items necessary for survival, they have no need to compete for these, although if we take present humans as models, they might still be driven to compete here. Let us suppose that their psychologies are such that they are not so driven. Implicit in the supposition that they are noncompetitive towards one another is their lack of envy, and their lack of desire to best others in their various activities. One further supposition: they are inclined to aid one another and to contribute to the overall benefit of their species.

There are instances in which a closely related word or phrase, substituted for a word or phrase about which controversy swirls, serves to quell debate. I propose to substitute the word "strive" for the word "compete." From some perspectives, there is all the issue between "competition" and "cooperation" that attends the political world-dividing controversy of capitalism versus socialism. Those holding a capitalistic economic worldview submit that it is the entrepreneurial spirit that has bestowed major gifts upon humanity, gifts that would be lacking if there were not individuals motivated to pursue beyond the status quo, and free to implement and market the results of their pursuit. Those favoring a socialist outlook argue that a better world results if the major modes of production are conducted for the benefit of all members of society, with distribution patterns emphasizing equality and minimizing wealth disparities. It seems prima facie obvious that a partial reconciliation of these views would be the understanding of a socio-political environment conducive to the equitable distribution of wealth, such wealth being created within a climate that encourages individuals and groups to strive to optimize various related goals such as; ease and delight in living, quality of mind, respect for natural and social environment.

Those who cannot forsake the spirit of competition will object that we become better in various respects for having overcome adversity. They will add that people not only ought to receive reward for their labor, but that if rewards are not commensurate with accomplishments, people's enthusiasm for accomplishment will be dampened.

There is much in the preceding paragraph to which I want to respond. It should first be noted that some of what is stated therein, although likely true for humans, is not likely true for their posthuman descendants as they have been speculatively portrayed. Humans of various cultures and societies harbor resentment when they believe others have been rewarded for achievements of comparatively less worth and difficulty than their own unrewarded work. Our speculated posthumans need not possess a reward/punishment inclination towards others, and they have already been ascribed a lack of envy. I agree that people often become better, in some intuitive sense of "better," for having overcome adversity. I add that this is also the case for our speculated posthumans. Since overcoming adversity necessitates striving, but does not necessitate competing, overcoming adversity need not be understood in terms of interpersonal competition. Posthumans must strive, individually and socially, if they are not to stagnate or devolve. That they must compete with each other is not a given, and in light of our conception of them it is prima facie believable that such competition would weaken, rather than strengthen, their living conditions.

It is not a simple matter to portray the average life of the average contemporary human, whether the portrayal is a lifetime summary or a description of a typical day. There is diversity even among members of the same society having similar social status. This remark carries over to the posthumans we have been considering. That they have similar abilities and similar exposure to various forms of knowledge, and that they live considerably more peacefully with one another than do humans, should not lead to the conclusion that their lives lack diversity. We can imagine that on a given day one posthuman might be engaged in musical composition while another is addressing concerns of chemical processing. The latter posthuman might be taking time to prepare meals, entertaining itself (note: not *her*self or *him*self in my speculation) in one of many possible ways, and returning from time to time to its chemical concerns. On the assumption that these posthumans have expertise in more than one area (recall that they lead indefinitely long lives), their labor, chiefly intellectual, will be varied.

I have indicated some of the general psychological features of the speculated posthumans. Their psychologies, cognitive and emotional, differ from ours to an extent that necessitates physiological differences between them and us. If they are sexless and if they live indefinitely extended lives, there will be further departures from our physiology. I leave it to readers' imaginations in what manner such differences are manifested, a constraint on such imagination being that these are comparatively near-term posthumans. Let us suppose that we are considering matters roughly one thousand years hence, rather than one hundred thousand years into the future.

It is impossible to know what will transpire with us. I have tried to sketch a possibility in general terms. I have not offered much specific detail regarding posthuman politics and posthuman social interaction. Given the speculative sketch I have provided, perhaps there are implications regarding these matters. I believe there are implications regarding their relationship to the environment, and to other creatures. I doubt, for example, that they will treat sentient creatures such as nonhuman mammals in a manner resembling the cruelty with which we often raise animals for food. To the extent that these posthumans take sensual delight in their nourishment, they will hopefully have restricted themselves to ways of producing foodstuffs that do not require raising sentient beings for slaughter.

Our current environmental concerns might be better addressed by these posthumans. They will have more control on world population levels regarding their species. We can imagine that new members will be created in roughly equal numbers to the members lost through internal malfunction or external impacts. Regarding such matters as energy production, heavy and light industrial techniques, transportation, and sanitation infrastructure, I prefer to say nothing—given that I have no clear ideas to which I can assign useful probability. I do have an idea about their agriculture, however, although I place little confidence in its realization, given the considerable range of alternative possibilities. They may not, after all, need to produce their foodstuffs agriculturally. Let us suppose that these posthumans still produce foodstuffs agriculturally, rather than in some completely synthetic manner. Let us further suppose that they wish to reforest much of the land that was deforested for agriculture. Given an abundance of energy produced with minimal environmental impact, they might construct a series of agricultural skyscrapers which use advanced hydroponic techniques. Their grown produce would be weather-shielded, and with multiple yearly harvests. Since to this point in time the essentially horizontal area requirement of agriculture has been instantiated on the earth's surface, much land has been deforested. If this surface area requirement is satisfiable by a vertical stacking of horizontal surfaces, much land could be reforested—if these posthumans prefer forests to farmed fields.

While I do not place much confidence in the above agricultural scheme being realized, I supply this scheme as an example of one of many ways that life as we know it might be profoundly transformed by our speculated posthumans. I chose the example of agriculture because, although there have been many ways in which agricultural procedures have been transformed since the initial human implementation of agriculture roughly 10,000 years ago, we still largely rely on the surface of the earth to produce our foodstuffs. But even as in our population centers we have built vertically to conserve surface area, so might our descendants "verticalize" their agriculture.

I have given much to these speculated posthumans. Some of what we humans have has been lost to them. They will not experience the joys of a sexually-charged loving relationship with another adult. They will not experience the joys of parenting. I believe that what they have been given exceeds what they have relinquished. I believe this summation remains correct even if a bit of what these posthumans have been given is subtracted from them. Perhaps they are not quite the renaissance individuals I have made them. Some might specialize in some areas of endeavor to the virtual exclusion of other areas, for example.

In any case, my assumptions in this chapter are idealized. Daily life even for these entities will not play on the frictionless surface I have tacitly assumed. There will be difficulties, with which I expect a society of such entities will be better equipped to negotiate. But weighing the joys and sorrows, and aware that I can only answer for myself, if asked whether I would want to be among these posthumans, my answer should be obvious to anyone who has read this far.

NOTES

1. This is the basis of Francis Fukuyama's strongest argument within *Our Posthuman Future*.
2. "Facing Immortality," *International Journal of Applied Philosophy* 19 (2005): 85–104. Reprinted with minor modification in *Inhuman Thoughts*.
3. Mill, *Utilitarianism*, 41–42.

Chapter 5

The Far Future

On the assumption that something of us survives into the far future, what will that something be like? This is a prima facie unanswerable question. Nevertheless, I propose to consider the question.

An initial task is to clarify the terms used in framing the question. The least troublesome such term is the phrase "the far future." By this I mean that time beyond any reasonable extrapolation of our current biological, psychological, and sociological circumstances. This elaboration is obviously vague, and to pursue the discussion in any manner one needs to set one's vagueness tolerance to a high level. Let us say that ten thousand years hence is far future, in that any prediction of what descendants of humanity might be then, including the possibility that there are no descendants then, is beyond any reasonable assessment. If ten thousand years is thought too close, allow upwards of one hundred thousand years.

"Us" refers to homo sapiens, modern humans in the strictly biological sense. In this sense, we have inhabited the earth for at least one hundred thousand years; possibly somewhat more. We likely originated in Africa a bit more than one hundred thousand years ago. These are the estimates as this chapter is being written. Comparatively superficial changes in height, skin color, and features, have to do with nutrition and breeding. Biologically, we are largely as we were when we migrated out of Africa. The manner in which we have changed psychologically is a more difficult question. To the extent that psychology reduces to neurophysiology, our nascent psychology remains largely unchanged. However, human neurophysiology changes as the maturing organism's brain develops connections, at least partially as a function of environmental stimuli. Given the large changes in both natural and social environment (I assume that a city-dweller's natural environment is the city) from the early days of homo sapiens, significant general disparities between a modern human's neurophysiology and that of an ancient human are to be expected. To the extent that human psychology does not reduce to neurophysiology, matters are complicated and unsettled regarding this question. Whatever further development the discipline of sociology undergoes, it seems uncontroversial that humans have undergone manifold sociological evolution.

Given that it is in some manner appropriate to consider the continuing evolution of humans as humans, the initial question is modified by the consideration of which "us" is the subject of the "something of us" that survives into the far future. If, for example, we largely retain our human characteristics (let us be intuitive regarding these), but extend average life expectancy to two hundred years—with attendant psychological and sociological alterations, then a somewhat different us than who we are currently is the subject of the question.

If we are projecting into the far future, then the point of worrying the referent of "us" is diminished, but not absent. As a species we might continue for some appreciable time, albeit with considerable modification. Yet considerably modified, we might—at least counterfactually—be capable of breeding with current human beings and producing fertile offspring. By certain standards we would yet be human. Taking our modifications into account, the question becomes rather, which "us" is partially surviving? We currently have psychological characteristics that reach back thousands, perhaps tens of thousands of years. That is, if we relax the requirement that psychology be as settled a science as the natural sciences and allow ourselves broad locutions, then we may say that are now psychologically quite a bit who we were. But perhaps who we are and were psychologically is not who we might be, while remaining biologically human by the above measure.

The point of this focus on who we are, psychologically, is that if we transform some of our general psychological characteristics while remaining human, then those future humans with their transformed characteristics might not imaginatively project themselves into the far future in the same manner as current humans. Individual differences aside, the similarities of the projections of these imagined transformed humans might bear marked differences with those similarities that would be found in the projections of current humans. For example, these transformed humans might have overcome the tendency of current humans to want a little more for themselves, over and above those adjacent. This tendency manifests itself in needless consumption among those who have discretionary wealth, as well as in impulses to outdo others. Indeed, current humans have valorized various forms of competition that their speculated successors might consider on balance deleterious.

Further, these successors might have expunged or mitigated certain of the so-called "reactive attitudes" that, while serving homo sapiens in their earlier survival struggles, have arguably outworn their usefulness.

Consider the various futuristic fantasies in literature and cinema of the past one hundred years. Almost invariably, agents who are recognizably human, with the strengths and foibles of current humans, dominate the proceedings. There are exceptions, of course. Typically these exceptions are humanoid individuals with exaggerated character traits that readers or viewers find either praiseworthy (e.g., pronounced even-temperament, enhanced cognitive abilities), or blameworthy (e.g., villainous characteristics, warlike). What is rarely encountered in these venues is a full society of humanlike entities with praiseworthy characteristics, together with an absence of human beings. Such is not the stuff of drama, although it might make for good speculative literature. As I am not a serious devotee of such literature, I may well have overlooked outstanding examples of what I say is rarely encountered. I doubt that I am mistaken in my assessment of the preponderance of futuristic fantasy literature, to say nothing of such cinema.

Modern humans have a tendency to celebrate well-known artists and scientists, although artists are more celebrated if their lives have had moments of turmoil. Haydn and Monet receive less personal attention than Mozart and Van Gogh. Similarly, scientists often receive public acclaim according to the startling nature of their discoveries and their supposed eccentricities. An individual leading a satisfied life of mind through scientific, artistic, or other cognitive behavior, while at least outwardly leading a placid, well-ordered life, attracts less attention. Yet a society largely composed of such individuals might, on reflection, be the sort of society most parents would choose for their children. Perhaps this is an unwarranted extension, but I submit that what one would choose for one's children is what one would project beyond one's children into the foreseeable future (i.e., the future in which inhabitants are not significantly altered as to biological, psychological, and sociological characteristics).

The beginning of the question to be pursued in this chapter is this: assuming a society such as mentioned in the above paragraph, what sort of projection, if any, might they entertain? It is an obvious oversimplification to presuppose a singular projection, rather than a plurality of such. Various groups within this advanced society of humans might have somewhat different projections. Indeed, the supposition that the limiting case of differing projections is that of each individual's projection overlooks the possibility, perhaps the likelihood, that an individual's thoughts on this matter will change from time to time. Granted, noting this possibility of an extreme plurality of projections, it must also be noted that many of these individual projections will be near-identical.

There are just so many ways that humans, even affectively and cognitively superior humans, can project themselves, either as projected individuals or as a projected species.

Would such humans project themselves at all? Suppose various current circumstances commonly termed "social ills" are largely abolished in societies of future humankind. Imagine a worldwide society largely purged of economic inequalities, with a widespread, high level of education (granted, what comprises such a level requires further detailing). The arts flourish, and the practice of them is well-subsidized. People are generally less prone to anger, political activity is more rational and less emotional, and understanding has largely replaced recrimination. People lead radically, but not indefinitely, extended lives, and typically lead them in good health, with universal access to updated medical treatment. In short, we are imagining that humanity, while still biologically human, has successfully completed what in previous chapters was termed a "transition."

I believe that were such a state of affairs realized, the still-humans would be seeking further transitioning. To recapitulate my reasons supporting this belief, the still-humans would seek ever extended lifespans to the point of indefinitely extended lifespans, mindful that such lifespans would in themselves likely require that they surrender their biological humanity. Although the brief sketch of the still humans' social circumstances indicates generally more peaceful lives than current humans enjoy, their lives are not without competition and disappointment. One source of such unease is the sexual attractions and tensions that, for biological humans, remain. As biological humans, there may be a limit to the lessening of tensions and conflicts arising from such attraction. Indeed, when all is done, if both sexes still exist then sexual difference still exists. If sexual difference still exists, it is difficult to imagine a human society in which such difference does not play some role in interpersonal relations.

As these speculated humans increase their longevity they will be aware of the diminished opportunity for children, given the increase in population. They will perhaps still view sexual activity as a form of intense pleasure, and as a means of reinforcing psychological intimacy. But in the passage of time, they may decide that sexual activity, or their being biologically sexed in itself, creates emotional distractions which outweigh its hitherto benefits. If they have sufficient control of their biology, they may choose to migrate from the biologically human to an unsexed species.

They may seek to implement additional changes for themselves, or their progeny. The many discussions regarding the so-called "problem of evil" provide a source for speculation. Those acquainted with theological and philosophical discussions regarding the existence of God in the face of human suffering and malfeasance are well aware of the sorts of imperfections of humans that are laid at the feet of Deity.[1] We are not interested here in theodicies, except insofar as they offer theologically neutral arguments for either the impossibility of alleviating particular human shortcomings, or the actual advantage of such ostensibly deleterious conditions over their removal.

In particular, physical pain is often cited as an unnecessary burden of humankind. There is, of course, much to be said about pain as a grouping of diverse psycho-physiological phenomena. This profusion of pain phenomena disallows quick universalizations regarding pain. Nonetheless, it is incontestable that the species of present humans would not have survived absent the existence of some physiological varieties of pain. The platitude of "nature's warning" cannot be fully dismissed. Those few individuals with congenital insensitivity to pain rarely live a normal lifespan. For example, we take for granted the physiological servo mechanisms that continually monitor our bodily positioning, so that discomfiture—a form of pain—results in readjustment prior to skeleto-muscular damage. Such does not happen in the case of people with absolute insensitivity to pain.

On the other hand, if consciously perceived physiological pain is indeed epiphenomenal, as I contended in chapter 3, then at least the logical possibility exists of having appropriate physiological responses minus the feeling of pain. However, it may prove physically impossible to engineer otherwise biological humans who have appropriate responses to danger-signifying stimuli while not having pain sensations. The invited question is this: if humans want to eliminate or substantially lessen various forms of physiological pain while not significantly decreasing survival probability, nor notably lessening desirable forms of life, ought they consider transforming themselves to some form of nonhuman?

Adding the multiform variants of psychological suffering to the discussion further complicates matters. Then there are those psychological states such as envy and ennui, which may not in all instances involve a conscious component of misery, but in some manner lessen various possibilities that might occur, but for their presence. Such possibilities are those that would ordinarily be considered positive, such as moments of creative thought. While it hardly needs adding that psychological torment that rises to the level of consciousness also impedes various positive possibilities, there are examples of creative accomplishment apparently motivated, at least in part, by such torment (note: given my epiphenomenalism I must allow that, strictly speaking, it is not the consciousness of suffering that is the causal motivator). The extent to which such suffering genuinely motivates these accomplishments is debatable, but for the purposes of this discussion this question need not be resolved. This is fortunate, in that our current understanding of these matters, including the conceptual schemes we employ in considering them, is arguably inadequate to the task.

Questions such as the value of suffering as a motivator for individual and social human progress can be left unresolved, inasmuch as these questions involve humans, and we are considering posthumans. That is, we are allowing ourselves the speculative luxury of imagining that various physiological, psychological, and sociological circumstances can be favorably adjusted. In this regard, it is helpful to examine some well-known statements indicative of what needs adjusting. First, there is Hume's character, Philo, from the *Dialogues*:

> Admitting your position, replied Philo, which yet is extremely doubtful, you must at the same time allow, that if pain be less frequent than pleasure, it is infinitely more violent and durable. One hour of it is often able to outweigh a day, a week, a month of our common insipid enjoyments; and how many days, weeks, and months, are passed by several in the most acute torments? Pleasure, scarcely in one instance, is ever able to reach ecstasy and rapture; and in no one instance can it continue for any time at its highest pitch and altitude. The spirits evaporate, the nerves relax, the fabric is disordered, and the enjoyment quickly degenerates into fatigue and uneasiness. But pain often, good God, how often! rises to torture and agony; and the longer it continues, it becomes still more genuine agony and torture. Patience is exhausted, courage languishes, melancholy seizes us, and nothing terminates our misery but the removal of its cause, or another event, which is the sole cure of all evil, but which, from our natural folly, we regard with still greater horror and consternation.

Next, there is Dostoevski's Ivan Karamazov:

> I want to embrace. I don't want more suffering. And if the sufferings of children go to swell the sum of sufferings which was necessary to pay for truth, then I protest that the truth is not worth such a price. I don't want the mother to embrace the oppressor who threw her son to the dogs! She dare not forgive him! Let her forgive him for herself, if she will, let her forgive the torturer for the immeasurable suffering of her mother's heart. But the sufferings of her tortured child she has no right to forgive![2]

Hume's Philo refers to long-lasting physical torments; toothaches, back pains, and so forth. To a sizeable extent, palliatives currently exist for such conditions. But various chronic pains still resist treatment, and practically everyone at some time endures agonizing, non-momentary pain. In the grip of such pain, subjects are ordinarily not at their cognitive or affective best. That pain inspires us to seek remedies is not contested. On the assumption that many such remedies are discoverable, whether the remedies are within human circumstances or are remedies by virtue of evolution to posthuman circumstances, the following points remain. Granting that humans have been motivated to discover ways of alleviating pain, these discoveries have been due to the exercise of cognitive powers. Pain is not the only motivator for such exercise, and when in the thrall of pain these powers are typically diminished. It might be that relieved of much of this pain humans would tend towards indolence. That supposition requires more than limited testing in controlled circumstances for its confirmation. One could perhaps as well suppose that freed of much pain, a notable increase of thoughtful activity would result.

Dostoevski's Ivan speaks of forms of suffering that are physiological pain and forms of suffering that are psychological torment, without clearly distinguishing the two. Nor need he, given his purpose. Still, Dostoevski's depiction of suffering as psychological torment is masterful. And it must be admitted that his portrayals of those in psychological torment, such as Ivan, are often depictions of the most thoughtful and creative of his characters. Biographical aspects of his life provide evidence that Dostoevski was personally acquainted with the suffering he depicts. Hence, unlike the case of physiological pain, there is not the easy position that we would be in all important respects better off without such suffering.

As Ivan Karamazov says earlier in the section from which the above quotation is taken, "there is suffering and suffering." Those suffering severe psychological depression, for example, are typically not capable of sustained thoughtful activity. The idea of the artist as a tormented soul is romantic; the Hellenic Greeks would not have easily recognized it (unlike their later Hellenistic counterparts). The picture of the suffering intellectual becomes less satisfying when held as depicting the intellectual endeavors of the natural scientist. I suggest that the image of suffering from which great art proceeds is misleading in the following manner. Outstanding artistic achievements often involve a balance between acquaintance and isolation. That is, the artist must be acquainted with what is happening in his/her field, but the artist must often isolate himself/herself during the creative process. Suffering of the sorts commonly ascribed to those artists who have been thought to suffer typically aids the isolation of the artist, and this isolation allows for what is original in his/her work. Of course, not all famous artists have suffered greatly. Some have led tranquil lives. It is difficult to imagine the artist suffering in his/her more creative moments. At these times, the artist is likely experiencing intense joy. Those habituated to the notion of artistic creation as a partial function of psychological torment hence can insist that this torment is minimally a partial cause of moments of creation, in terms of relief consciously or subconsciously sought by the artist.

Some suffering is caused by external happenings, and some is apparently internally generated, in a manner not commensurate with publicly observable circumstances. If we can build a better world, we can do much to alleviate external causes, assuming part of what is meant by "better" in this context is the alleviation of these causes. Similarly, if we can build a better human—perhaps by transforming human selves into posthumans, we thereby end much internally generated suffering. On these assumptions, the loss of artistic creative activity is debatable. As has been noted, not all artists have shown pronounced emotional torment. Perhaps there is a kind of art that demands torment. It seems likely that if much suffering is alleviated, then suffering will fade as subject matter of artistic depiction. Further, art that consists in the expression of inner torment will not remain, absent such torment. While I do not completely endorse Plato's overall attitude towards the arts, I voice the following Platonic sentiment: if the price of a world with limited suffering and much possibility of intellectual development—including various forms of aesthetics, perhaps adding some hitherto unknown forms—if that price is the loss of some current forms of art, then to my mind the price is easily paid.

Not everyone will agree with this last thought. I request that those in disagreement introspect their images of notable intellects in various other fields of thought. The natural scientist has already received mention. Important figures in other fields are easily brought to mind. Not all these figures are unmarked by observed or self-recorded suffering. But the remarks above are especially applicable here. In most cases of this suffering it is less difficult imaginatively to detach the suffering from the thinker than in various instances of artistic endeavor. Even if much art is to be surrendered, and I don't see the necessity of this, much opportunity of creative thought remains, and under happier circumstances.

Some final notes to this discussion. Notions such as suffering and torment belong to an interrelated vocabulary that has had its uses up to the present, but might not carry into the future. In present usage, largely all humans have had moments of misery, sorrow, and anguish. These related terms should be distinguished from terms such as restlessness, vexation, and dissatisfaction, also had at some time by most humans. To whatever extent the former set of terms and the latter set of terms carve psychological reality at the joints, or perhaps are supported by or indicate underlying concepts that accomplish this, these two sets should be separated from each other. People, or those evolved from people, can get along with diminished levels of suffering and torment. However, a future in which humans or posthumans are not challenged by problems, and hence vexed and dissatisfied, is unattractive. Indolence in the short term can be restorative. Long-term lassitude is not satisfactory for developed cognitive beings.

I have discussed pain and suffering at length for the reason that one powerful motivation for transitioning to something other than what we currently are is the avoidance or mitigation of pain and suffering. It is nevertheless possible that with various pain-relieving modifications we could remain human, at least in the sense that those so modified could interbreed with those unmodified and have fertile offspring, although the pain-lessened status of such offspring at birth is problematic. When other, previously discussed modifications such as radically extended lifespan and sexlessness are added to lessened pain and suffering, however, the likelihood that entities having all such characteristics are human in the biological sense of the term is diminished.

Technological innovation is generally predictable only in the near-term, and only by those who follow developments closely. Much of my speculation here relies on scientific and technological advances, together with those human and posthuman events which motivate these advances, and are influenced by their appearance. Hence, inasmuch as some of my speculations reach into the far future, such speculations concern unpredictable circumstances. Aside from whatever enjoyment is derived from their contemplation as fanciful possibilities, their practical value resides in the effect they have on those current sensibilities of ours that partially determine our manner of regard for one another, and for ourselves. If we imaginatively project ourselves as largely cognitive, creative beings in cooperation rather than competition with one another, such thinking might have some influence on our current behavior.

In my earlier work, *Inhuman Thoughts*, I speculated on both human and posthuman improvements and transformations. Some of these improvements and transformations have been mentioned in this work, including an increase in our cognitive abilities, both as humans and posthumans; radical extension of lifespan; sexlessness; and alterations of our emotional constitution. In my previous work I included some near-term and far-future speculation not discussed in this book. They are worthy of mention here, in that some of them fit into the category of medium-stage–far-future speculation, thereby functioning as a transition to temporally and philosophically deeper speculations.

There is one possibly nearer-term speculation from my previous work, not mentioned above, that also fits into the category of medium-stage–far-future speculation. Regarding this speculation from my previous work, therein I considered the possibility of what I termed "parallel consciousness." Given my epiphenomenalism, stated in chapter 3, I will avoid the term "consciousness" in my summary of this notion. In simplest terms, the idea is that a locus of cognitive activity, human or posthuman, has the ability to pursue simultaneously more than one cognitive task. In other words, the cognizer *multitasks*. For example, the cognitive agent might simultaneously be at work on a problem in theoretical physics while composing a poem. Although humans are capable of what are commonly termed "attention shifts," tasks such as just described seem beyond current human capabilities—provided that the word "simultaneously" is given due regard.

In my previous work I considered such cognitive parallelism to be a posthuman possibility, although I entertained the thought that humans could develop this capacity. The means by which humans might develop this ability include various possible "thought-exercises" (the success of such techniques in achieving significant results is doubtful), or what is more likely, some form of directly-targeted neurophysiological modifications. These modifications might be such as to allow for the continued use of the term "human." It is left to the reader's imagination to consider possibilities here.

Another speculation from my previous work was that posthumans be endowed with more cognitive, even theoretically cognitive, sense modalities. I focused on vision in my discussion, but allowed for similar development of other senses. My overall intention in this discussion was to lessen the current divide between perception and theoretical thought. Vision might not only be developed as to bandwidth (extending above and below what is currently termed "visible light"), but might be cognitively enriched in various manners. For example, assuming that a posthuman has something akin to a visual field, that field might contain "readout" information somewhat like the information currently available through the viewfinder of various optical devices. This information might include more than distance, intensity, and so forth. It might include frequencies, wavelengths, phase angles, and other theory-laden aspects of the radiation detected.

Although I mentioned the possibility of an asocial existence in chapter 1, I did not comment extensively on it. In *Inhuman Thoughts* I considered an asocial existence for posthumans. I did not mean that these speculated posthumans would not communicate with each other, or not depend on each other. Rather, I intended that these posthumans might not have need for the sort of casual contact with each other that most humans regularly need, on pain of psychic distress absent such contact. Of all my previous speculations, this was perhaps the most lighthearted. The serious point of it was that, unlike most humans today, who praise thought but do not regularly engage in it, our posthuman descendants might welcome lengthy, uninterrupted thinking, perhaps done in solitude.

I believed that these speculated transformations would be in the far future. I am less convinced of that now. I never believed in the likelihood of any of these speculations. That is, I believed then and believe now that all such speculations suffer the common fate that the near-distant future, let alone the far future, affords more novelty than can be appreciated from the present vantage point. What I believed then and am somewhat doubtful of now, is that these speculated transformations, or others of a similar degree of novelty, would not happen in the near-distant future, and so were relegated to the far future—say ten thousand or more years hence. I believe now that such transformations have a likelihood of occurring earlier, perhaps within the next thousand years. This belief, and my earlier belief, both assume that humanity (or transitioned humanity) comes to a prosperous "golden age" within the next few centuries, and from there undertakes radical changes. Again, I do not think my speculated changes are likely. But rather, that some such changes are likely, accepting my optimistic assumption of a golden age. While the notion of "some such changes" is vague, possibly vague beyond redemption, I am hopeful that it allows for an imaginative vista of a spread of possibilities at an intuitively sensed degree of difference from the present. It is of course possible that my optimistic assumption is unfulfilled, and that in its place dire-to-horrific events befall our descendants.

Granting the weakness of all such speculation, I want to press further. The possibilities thus far entertained in this work and my earlier work have tacitly allowed that the bearers of these changes display an overall humanoid physiognomy. Features of such form include bipedality, vertical axis left-right symmetry, horizontal axis and front-back asymmetry. Much science fiction cinema adopts this description in its depiction of intelligent, nonhuman creatures. The evolutionary success of homo sapiens provides a respectable argument for the advantages of such physiognomy, and there is no shortage of post hoc explanations of the survival value of our human configuration. The combination of comparatively large craniums with savanna dwelling hunter-gatherer living circumstances is said to be well-served by an upright posture, enabling superior distance vision. Further rationales are given for other humanoid features.

If we continue the optimistic assumption given above, the plausibility of our posthuman descendants gaining extraordinary understanding and control of their biology is strong. They will have reason to consider critically their continued acceptance of humanoid form. Such form served humanity for the majority of its time. It will likely serve posthumanity in its transitional and early post-transitional stages. It is a delicate question as to the extent it will continue to serve into the far future. To some extent this question is independent of the question as to whether posthumanity will be wholly or largely biological. The all-purpose robots we imagine often have humanoid characteristics, although the match between our casual imaginings and serious implementations, if such is possible, may be wide.

I am aware of the repellent, frightening effect of hinting at the possibility of transformation of humanity to a non-biological posthumanity. I believe some of the repugnancy attending this suggestion is based on our current view of those non-biological entities capable of performing a limited number of functions previously the sole province of humans. Such entities, robots, are currently incapable of the broad-spectrum cognitive competencies of humans, and they lack genuine affective characteristics. This latter lack is problematic, especially given my epiphenomenalist position. I do not believe emotions are completely accounted for by behavioral dispositions. Rather, I believe that underlying brain activity accompanies the "felt" qualities associated with emotions, and it is this brain activity, together with various behaviors and various conscious experiences, that constitutes what we consider the phenomena of emotions. At most, some current robots manifest some behavioral aspects of human emotions, but not the rich, complex, and currently ill-understood brain activity of humans, which activity causally underlies both the behavioral aspects of humans and the quality of human emotional phenomena.

There are gains in evolving to a non-biological existence. The fragility of biological configurations might be largely overcome if human or posthuman biology is superseded by embodied beings with more durable constitutions. Of course, there is much to consider here, even allowing the lightheartedness of such speculation. If reasons otherwise incline towards remaining biological, there are various future possibilities allowing for continuity of an individual in the face of unexpected impact, internal or external, to the individual's biological self. If, for example, the individual's genetic program is on file, and if the individual's mind/brain is in regular "backup," then there is the possibility of restoring the individual with only some loss of continuity, which might be a loss of minimal duration.

The complexity of biological systems, at both the macro and micro level, cautions against an ill-considered replacement with non-biological surrogates. The time may come when various familiar and significant human anatomical parts are universally replaced by non-biological surrogates. Prosthetic limbs have a long history; artificial hearts are recent. At this point, total replacement of brains by non-biological components is beyond adequate conception. Ten thousand years into the future, such replacement might seem routine, even as in recent memory computers have replaced various clerical workers in the performance of various clerical tasks.

I have been focusing on the question of retention of biological composition by our far-future descendants. This is arguably a more profound concern than the question of their overall physiognomy. Still, it is intriguing to remind ourselves that our far-future descendants might not resemble us in their outward form. Evolution has long bestowed the hominids with the forward-facing eyes of predators and other creatures whose survival has been aided by concentrated binocular vision, rather than the panoramic vision of those prey to carnivores. Absent the environmental circumstances from the Pleistocene to the present, with little information at present with which to determine far future environments, we might nevertheless speculate regarding such distant environments. A constraint on such speculation is that it be done in the spirit of uncertainty.

Rather than speculate regarding what environments our far future descendants might face, however, I would like to speculate on various matters of their physical constitution. Their sensory apparatus will tend to reflect their environmental possibilities. Perhaps they will have high-bandwidth, omnidirectional vision, which necessitates something functionally akin to the current visual center in the human cortex, except that it will be modified to accept and integrate this expanded input. Vision so altered might better fit an environment in which neither flight nor pursuit is typically needed. Perhaps given their environmental circumstances, a different form of stance and locomotion than bipedalism/walking is favorably indicated. If the great majority of surfaces on which they travel, unassisted by devices external to themselves, are smooth, they may have forsaken lower limbs in favor of roller-like appendages. They might have a variety of appendages that they somehow manifest on demand, those temporarily unneeded being retracted in some manner.

Likewise, the internals of the far future posthumans might substantially depart from those of current humans. This is of course to be expected if they fully shed their biological aspects. Given likely scientific advances, these posthumans probably would be capable of substantial re-engineering of their internal biological characteristics, should they retain their biological constitutions. We can imagine that various breakdown points of current human biology will be reinforced or replaced by significantly less failure prone surrogates. This contemplated strengthening is in addition to whatever scientific breakthroughs enable control of the aging processes.

Assuming that various transitions occur among the far future posthumans, there are questions as to the manner in which these transitions are managed. Do they occur more or less simultaneously among the population, or as desired by each individual? Are they uniform, or varied? Is there oversight, and if so, what is the nature of it? Is there anything resembling political debate in the process of adopting new characteristics?

Of course I have no answers to these and related questions. But I have an expectation. If our descendants survive and prosper in the general manner I have been indicating, then I expect they will make their decisions with what would appear to us to be unprecedented wisdom. They will have a more solid grasp of complexities, and thus of what outcomes of decisions can be projected and to what extent. They will make their decisions carefully, guided by an ethics more in touch with the deep realities of the internal and external factors of their situation. It almost must be like that for them, because if it is not they would never have gotten that far.

My expressed attitude in the preceding paragraph is in direct opposition to a well-known (apocryphal?) statement of Einstein's, and is illumined by comparison. Einstein supposedly responded when queried as to the manner of weaponry for a third world war that, although he did not know how World War Three would be fought he knew how World War Four would be fought—with sticks and stones. A sobering statement of human devolution, to which I counterpose an optimistic possibility; posthumanity finds its transitions in possibilities that the most learned and noblest of us, such as Einstein himself (at least as we represent him), can only dream.

To this point I have considered far future posthumanity as a plurality of individuals. Assuming there is a developed, far future posthumanity, these supposed individuals may have at some point united to form some manner of single being. What that manner of being is invites speculation constrained, if at all, by fundamental physical constants. If, for example, this individual being is widespread (say, over the entire solar system) then the speed of light limits whatever communication occurs among its parts. Current computational machinery is limited by the speed of light, in that light travels short distances in intervals less than a billionth of a second, yet current computer clock cycles are less than one billionth of a second. Hence, internal communication among spatially separated parts of a current desktop computer is constrained to close distances, if it is to occur within a single cycle, or else is allowed to occur over more than one cycle if the distances exceed several centimeters.

The reasons that would persuade the set of far future posthumans to surrender their individuality for a unity of some currently incredible sort are generally the stuff of science fiction. They might be compelled by matters of security against an external threat, be the threat some sort of life form, cognitive or noncognitive, or be the threat lifeless, but dire. Perhaps their individuality would not be wholly abandoned. Rather, they might enter into an arrangement of interlinked minds, from which a unified mind emerges on such occasions as needed, while at other intervals their individual selves are active. A somewhat applicable analogy is that of a computer network, interlinked in such a manner that in appropriate circumstances, or perhaps on demand, the network alters its behavior from linked member units to one large unit, formed by the combined computational activity of the smaller units operating in concert under some unifying program.

Among these and other possibilities there is hopefully the following constant. What evolves and manifests is in pursuit of knowledge and nondestructive creative possibilities, and what so pursues has some sort of consciousness of its activity and the results of it, and this consciousness is at least partially constituted by some sort of enjoyment. I am aware that this hope is expressed from my perspective, by which I do not mean my solely individual perspective, but rather what I take to be the highest perspective of my species in its current historical positioning. I have argued for this perspective in the previous chapter. It would be blindly arrogant of me to suppose that everyone accepts my argument. Those who do not accept my argument may take some comfort in the following. Even granting the correctness of my assertion regarding current humanity's highest thoughts, it is beyond establishing that these speculated far future descendants will share these values. Consciousness, creativity, and cognition may come to mean little for them. Against the insistence that these speculated posthumans could not maintain and advance if they did not possess at minimum powerful cognitive abilities, it may be replied that they have provided for themselves in such a manner that background systems ensure their welfare, leaving them to their preferred modes of existence.

From my personal and, hopefully, species standpoint, it is difficult to consider an existence of reduced cognitive activity for such beings. Such difficulty does not constitute an argument against the possibility, or likelihood, that these beings will prefer, say, a largely thoughtless existence of indolence, or even an existence lacking consciousness. I cannot imagine that evolution, natural and artificial, would lead to this, but again, what I can and cannot imagine does not settle this difficulty. Humans have values. Speculated near-future posthumans will almost certainly have values. Optimistically considered (from my perspective), near-future posthumans will have refined values similar to those of the humans they have replaced. Given further transformations, the most that can be stated is that if such beings (or being) continue the cognitive quests initiated by humans, they will likely retain various human values, although they will have considerably refined them.

Whether far future posthumans remain a plurality of individuals or somehow meld into one individual, their identity over transitions is a problematic matter. For mortal humans, such identity beyond biological death is problematic only for those believing in some sort of survival of quasi-psychological spirit beyond biological death. For those not believing in such survivability of spirit or soul, and I count myself among them, biological death ends whatever self-identity one has, at least to this point in human history. Speculatively, human matters can become more complicated. If at some future time humans are able to store and periodically update their psychological characteristics in significant detail, and an appropriate sort of body-replication is available, then it is possible that individual human self-identity can survive biological death. On fulfillment of the preceding assumptions, such survival would be a matter of appropriately downloading the most recent update of an individual's psychological characteristics into their replicated body.

As noted in the chapter "Minds and Related Matters," within contemporary analytic philosophy there is a large body of literature on the topic of personal identity. I do not intend to enter the discussion here in any sustained manner. Rather, the following remarks will need suffice. As preface to these remarks, it is noted that the notion of transitioning in this discussion is at best suggestive. I am unable to supply useful criteria regarding the point at which humans transition to posthumans. Given sufficiently radical alterations, there will be large, intuitively based agreement that what has evolved is no longer human. But again, if these alterations are preceded by a number of stages, there may be disagreement as to which stage marks the transition. For any speculation regarding various alterations among posthumans, what constitutes points of transition will similarly be open to deliberation and debate.

I believe that whatever the state of posthumanity, assuming upward progress from humanity—where such positive direction is to be measured by a still-to-be-developed ethical theory—posthumans, be they a plurality or a unity, will typically seek to preserve something of themselves through their transitions. I base this belief on the position of the being choosing the transition. I am assuming that far future transitions among posthumans will not be accidental, but will be chosen. Hence, the chooser will likely want to be that which has transitioned, rather than a being wholly dissociated from that which has chosen. I realize that I may well be anthropomorphizing that which is beyond the anthropomorphic; projecting a human's motivations onto that which is not human. But if these beings are lacking in conatus, then the question of their continued survival in any form seems appropriate.

As indicated in the preceding paragraph, I am not fully confident in the reasoning behind my assertion of continuity of self-identity among the posthumans. I am doubtful that the sense of self-identity that survives an immediate transition will persist through a multitude of transitions. Because humans and all heretofore sentient earth creatures have limited lifespans, they cannot serve as examples of such gradual loss of self-identity through transitions of the kind here contemplated. And it may be that the sort of cognitive powers possessed by the posthumans, and never surrendered—only improved—will allow a complete self-identification all the way back to those humans who transitioned to beings with indefinitely extended lifespans. Alternatively, there may be moments at which, under internal or external threat, it behooves the posthumans to transition without preserving self-identity over the transition.

NOTES

1. A good anthology of pro and contra views on this issue is Nelson Pike, ed., *God and Evil* (Englewood Cliffs, New Jersey: Prentice-Hall, 1964).
2. Fyodor Dostoevski, *The Brothers Karamazov*, trans. Constance Garnett. The quotation is from the "Pro and Contra" section titled "Rebellion."

Chapter 6

At the Limits of the Conceivable

What will engage our posthuman descendants (or descendant) roughly one hundred thousand to one million years hence? Assuming, of course, that there are such descendants, that they are posthuman, and that they have prospered in manners indicated in previous chapters. We are obviously at the limits of speculation; minor details should not be considered. Whatever is said beyond the widest generalities cannot be meaningfully evaluated as to probability, and wide generalities themselves are difficult or impossible to evaluate. Still, there is some worth in taking our speculations to the limit. That which limits us might be overcome by these descendants. At the least, such speculation will allow us an interesting view of these limits.

Some such limits and their overcoming have a history with which we are acquainted. At the end of the nineteenth century both special and general relativity, together with quantum mechanics, were beyond the powers of conception of most then-practicing physicists. Physical phenomena were thought to be fully determined, and some of the ideas of relativity theory, such as the non-Euclidean nature of physical space, and the constancy of the speed of light in all velocity frameworks, would have been thought by many to conflict with fundamental principles of logic and mathematics (as indeed they were thought by some, on their introduction). In the history of human overcoming of presumed ontological and epistemological limits, these are but recent examples. There have been many more.

A person or team of persons might have overcome a presumed limit for reasons of vanity, curiosity, or for further expected benefits. I assume that if these (or this) speculated deep future posthumans overcome current limitations, it will be for the latter two of the three given reasons. I assume that whatever physical reality underlies the superficial notion of vanity will be long departed. In terming the notion of vanity "superficial," I am not saying that notions such as vanity, pride, and envy have no useful place in human commerce as it has been and is likely to be for somewhat longer. Together with a multitude of common psychological and sociological concepts, these notions found common descriptive and explanatory sketches of various patterns of behavior.

A person might be described in terms of their brain activity, their appropriate environmental history, or whatever deeper-level concepts support a more thoroughgoing account of their behavior. That such deeper-level concepts will coincide with the more common psychological and sociological notions commonly employed is less than assured. Perhaps it is an open question whether descriptive-explanatory items such as acting out of vanity will end up in similar fashion to concepts such as phlogiston and caloric, or will rather be encompassed by more comprehensive theories, as pre-quantum chemistry with its ions and valences was encompassed by the quantum theory of the 1920s.

What I am assuming is that notions such as vanity, pride, and so forth remain legitimate, higher-level descriptive, if not explanatory, concepts. The difficulty here should be apparent. I am attempting to speculate on deep future beings using terminology that might be outdated in the deep future, even if it is genetically human beings who populate the deep future—although I doubt they will. If these concepts do not remain legitimate, then acting out of vanity cannot be correctly ascribed to future beings, and retrospectively has been mistakenly applied to humans. Alternatively, if acting out of vanity remains an acceptable description of some forms of behavior (and perhaps "behavior" is itself a suspect notion), whether of humans or some other type of being, then I do not believe that the sort of posthumans about whom I am now speculating will act out of vanity, or envy, or pride, or greed.

There are possibilities for these speculated beings that are at the edge of conceivability, and there are possibilities that test the notion of conceivability. These possibilities generally are motivated by cognitive urges that I attribute to these beings. That entities so far removed from us, temporally and otherwise, would have such urges is problematic. It is not difficult to assume they do not know everything, although I must grant that the phrase "knowing everything" is philosophically problematic. It is further granted that if an agreeable sense can be given to the idea of knowing everything (there may be logical difficulties here), then it seems not beyond possibility that they do know everything. If they know everything, then they will likely not be cognitively motivated to know more. Relaxing the supposition that they know everything, and avoiding the notorious philosophical difficulties of the term "knowledge," it is less contentious to assume that their cognitive capacities greatly exceed those of their distant human ancestors. Perhaps they have reached a point of satiety, and are no longer cognitively active.

It is of course impossible to state what activity these speculated beings will pursue, unless one chooses to assume an activity as part of the speculation. I thus assume that they will engage in cognitive activity. Examination of various fields of cognitive activity, which I speculatively assign these beings, will constitute the discussion of this chapter.

Suppose that our deep-future descendants have access to powers that enable them to examine and recreate nature, locally or globally (let us rest intuitively on these last terms). Such speculation is not as far fetched as might initially appear. Had the Superconducting Super Collider been constructed, for example, physical conditions approaching those at an infinitesimal interval after the supposed Big Bang might have been realized.

Let us make some fanciful and not-so-fanciful assumptions. Assume that the Big Bang is a generally correct account of the physical beginning of our physical universe (I say "*our* physical universe" to leave open the possibility of other, in some sense "co-existing" universes, as posited in "brane" and "bubble" theories). Assume that so-called "universal constants" (e.g., Planck's constant, speed of light in a vacuum) are "universe-relative," and those constants of our universe emerged shortly after the Big Bang. Assume that our speculated descendants have devised portable apparatus for local recreation of the Big Bang singularity; that is, somehow Big Bang conditions, or pre-conditions, are minaturized to the extent that they can occur, on appropriate demand, within a restricted domain of the physical universe.

Some of these suppositions may be impossible by present lights. In any case I am beyond my competence, and relying on various sources of varying reputability. It may be questioned why actual, rather than simulated, recreation of Big Bang conditions are of greater cognitive value. It seems credible that in this speculated deep future something akin to present day computer simulations exist, although of immensely improved capabilities. I agree that modes of simulation, rather than actual physical exploration, may be the cognitive tool of choice in the deep future of my speculations. I note further that notions of "actual physical exploration," although generally well-understood currently, might acquire significantly altered meaning in this future. Some of what is said below will serve to elaborate this thought.

These deep future posthumans might choose to recreate Big Bang conditions locally, rather than simulate such conditions, for a number of reasons. It is possible that the recreation is easier than simulation, to the degree that simulation is not possible whereas recreation is comparatively within reach. Consider that various hydrodynamic (fluid dynamics) circumstances are currently not realizable in detail as computer simulations, but are sometimes as easily realized physically as turning on a faucet. At times, the complexity of the phenomenon of interest is overwhelming, whereas the overall phenomenon is easily created.

Another reason our deep future descendants might prefer realization to simulation regards their possible wish to be involved with the realization. For example, recreating various extreme physical conditions might gain them access to phenomena otherwise unattainable. Perhaps there are alternative universes which, given our present circumstances and abilities, cannot be apprehended. Our deep future descendants might discover "paths" to these alternative universes, and be able to "travel" these paths without harming themselves, and with the possibility of return.

These speculations face the criticism that at least some of them are incompatible with physical constants and laws of nature. Local recreation of Big Bang conditions apparently demands an infinite or near-infinite amount of energy, for example. A Big Bang singularity would exceed conditions necessary for black hole creation. Further, if such a singularity could be in any sense *locally* created, in what manner could it be controlled so that it did not in effect "erupt" into another universe within the extant containing universe? I will address this concern below.

Surely, from our present stance these speculations cannot be evaluated as to probability or even possibility. They are not simply at the limits of what is conceivable. In some senses of the term "transcend," they transcend these limits. We are already acquainted with examples of transcendence in terms of mathematical and physical singularities; such singularities being circumstances in which discontinuities of various sorts are positioned within an otherwise "well-behaved" context. For example, the relevant mathematics of subsonic flight indicates a singularity in the transonic region, although the reality of supersonic flight (or that of various ballistic projectiles) shows the need for correction factors which describe a less than infinite resistance to bodies approaching the speed of sound. If various classical equations have at some historical periods determined the limits of what is physically conceivable, some of these ostensible limits have been transcended.

In the speculated case of "forcing" singularities through the creation or recreation of extreme physical conditions there are additional considerations. Most notably, the results of such recreation might include an altered set of physical constants and laws. It is one matter to entertain that the constants of nature do not have the status of logical certainty. Such a view is frequently attributed to Hume. Whatever Hume's actual views on this, he occasionally seems to ascribe an underlying necessity of some sort to observed regularities. In the case of universe-making initial conditions, however, it is an additional assumption that the resultant "cooling" will yield a cosmos lawfully similar to the one with which we are presently acquainted.

Among the questions raised by the speculated possibility of universe-making is that of the existence of meta-laws. Such laws, if they exist, are trans-universal. That there are such laws is as at least as much a matter of conjecture as that there are, actually or possibly, alternative universes. A further question regards the form of such laws. If they exist, can they be expressed mathematically—in any intuitively recognizable sense of "mathematical expressibility?" If not, in what sense can they be understood as laws? I will return to this question. But we should note that it might be the case that there are actual or possible alternative universes (possible but not actual in the event that ours is currently the only universe, but others can be created) and no trans-universal laws covering phenomena in all universes. Either the creation or the pre-existence of alternative universes would then have some degree of randomness to them, however deterministic the phenomena within them might be. We should also note that the sentences immediately preceding allow for the possibility that the mathematical systems from which judgments of randomness are intelligible might be restricted to our universe.

Let us return to the question of our present universe containing the effects of an experiment wherein universe-making conditions are evoked. If it is assumed that universe-making conditions lawfully lead to the creation of universes, then apparently there are some trans-universal constants. This assumption need not be made. In the largely unconstrained atmosphere of these assumptions, it can as easily be assumed that conditions responsible for our universe do not necessarily lead to creation of other universes—assuming that it is possible to localize such conditions within our universe. If indeed multiple universes have been, or can be, created, perhaps it is a matter of chance whether similar initial conditions of this sort have similar effects. Another possibility is that there is a trans-universal principle regarding the creation of universes, but the "cooling" or congealing of initial resultants into material packets is not governed by trans-universals, nor are the subsequent physical constants that arise upon such congealing so governed.

Whatever the case, it is the speculative supposition here that at some point in the deep future our posthuman descendants (or perhaps unified, singular descendant) might be in a position to investigate questions that are beyond our current ability to investigate. Universe creation, even if limited in scale, does not seem a matter to be undertaken lightly, at least from our perspective. If our posthuman descendants are capable of doing this, they will hopefully for their sake have considered the consequences, including the immediate consequences of danger to themselves in light of the forces that will be liberated. These descendants may well be of significantly altered form than that of present humans. They may be able to alter their form to something other than matter, let alone biological matter.

We have been reconnoitering the far-fetched deep future in these speculations. I cannot claim any likelihood for the matters upon which I have speculated. I do not want to place them beyond all likelihood, however. In the span of one hundred thousand to one million years from now, that to which present humanity might evolve is not at all predictable. Perhaps these speculated posthuman descendants will be capable of re-creating their entire physical universe. On the assumption that there are other sentient beings therein, one hopes that all attempts are made to seek these other beings' agreement, or at the least consider their situations, prior to undertaking such action as re-creating the physical universe in which they have evolved. If our posthuman descendants have universe-making and universe-recreating power, and do not know the situations of other sentient beings, they will hopefully exercise proper judgment and refrain from using such power.

Thus far we have speculated on possibilities of deep future posthumans at what we regard as cosmological limits. It is possibly mistaken to assume that all or even a majority of their activity will be cognitive, and that it will be at these limits. Focusing on the cognitive at these limits, however, epistemological and ontological questions are raised regarding the conceptual instruments generally assumed to underlie such activity; those of logic and mathematics.

Those familiar with the philosophical discussions regarding the foundations of mathematics, the nature of mathematical and logical truth, the nature of necessity, and so on know that there are problematics associated with these concerns. Questions such as the nature of numbers, the ultimate grounding of inference, the appropriate semantics for propositional logic, are only the more general of a host of further particular issues discussed by those interested in looking behind the apparent certitudes and powerful applications provided by mathematics and logic.

Some of these issues arise within the formalisms of mathematics and logic. There are notoriously unsettled questions regarding infinity, continuity, and formal paradoxes of set theory. Other questions concern the interpretations (semantics) of these formalisms. The literature regarding these questions is large and increasing.[1] While I cannot engage in a sustained discussion of any of the issues in this context, I want to sketch and casually explore some logico-mathematical considerations related to the extreme physical circumstances presented above.

If one is some sort of ontological Platonist regarding logico-mathematical entities, then one likely maintains that the nature of such entities is unrelated to physical circumstances (be it noted that the recognized use of the term "Platonist" in this context is not necessarily tied to any position held by

Plato). Not everyone countenances this form of being for these entities. The contrary positions of formalism and intuitionism are well-known alternative positions. These alternatives possibly entail a view of mathematics and logic that is to some extent dependent on physical circumstances. In maintaining that the natural numbers are fundamentally human (or perhaps more generally, sentient) intuitions regarding the passage of moments of time, the intuitionist account of the natural numbers relies on a notion of time that might not be appropriate were physical circumstances radically altered in manners suggested above. Similarly, formalists claim that logic and mathematics are specialized formal games involving specific rules for the manipulation of uninterpreted signs. There has been a fair amount of philosophical attention paid the notion of rule-following, with some focus on the employment of this notion in formalist philosophies. I believe there has been less attention given the nature of the signs themselves, within the general context of discussions regarding formalist philosophy. Unless one is a Platonist regarding the nature of these signs, these signs have their being as a spread of physical possibilities, with many of these possibilities having vague boundaries. For example, it might be a question of interpretation whether a vertical stroke on paper in a particular context indicates the number one or a lower case alphabetical letter typically preceding the letter "m." There are, of course, a myriad of manners of indicating the number one, but many—if not all—these manners have some sort of physical basis. Once again, in an environment (perhaps a universe) of radically altered physical nature, the physical basis of formalism might be in question (assuming the correctness of my belief that formalisms are physically based).

The remarks regarding mathematical entities in the preceding paragraph apply also to the entities of logic. Of special concern is the fundamental relation of logical necessity. Once again, there are varying positions with reference to the ontological status of the relation. The two poles which serve as rough attractors of these positions are those of *de res* and *de dicto*. The former position is ontologically realistic regarding necessity; there are necessary relations "out there," whereas the latter position considers necessity as our decision to regard symbols in a certain manner, and hence may be regarded as a variant of nominalism. These characterizations are loosely presented here, and there is much refined discussion of these matters in the relevant philosophical literature. While I find it difficult, if not impossible, to surrender my Platonistic outlook towards this fundamental relation, I accept that in a radically altered set of circumstances—which I can neither imagine nor coherently conceive—this relation might not exist.

In taking these speculations to what I believe are their limits, I recognize various philosophical difficulties. I have been suggesting physical circumstances in which such universal notions as unity, plurality, necessity are not applicable. Put differently: logic and mathematics, as generally understood and philosophical differences aside, do not pertain to these circumstances. This suggestion is not novel. In the first half of the *Theaetetus*, Plato explores the notion that the underlying condition of reality is constant change (152d–157c, 179c–183b) and concludes that such a metaphysical hypothesis is self-refuting, in that it disallows the motionless (non-fluctuating) reality of the thoughtful speech that frames the hypothesis. As Plato appears to argue, a rationally intelligible reality necessitates that not all things are in constant change. More particularly, as Plato emphasizes, what allows for the rational intelligibility of perceived reality is the fact that not all perceivable (or, as we might say, physical) things are in constant change in every manner of change (181d–183a). If Plato is correct on this point, it follows that if all things are in constant change, in every manner of change, our systems of rational intelligibility—logic and mathematics—are undermined. More generally, the truism is suggested that if physical reality is altered beyond present conceivability, then modes of present conceivability are inapplicable in such circumstances.

Of course Plato does not endorse the possibility that has been speculated here. Further, he might well challenge the possibility of this speculation, in that he likely accepts that possibility entails conceivability. Hence, what is not conceivable is not possible. This train of thought places some burden on the term "conceivability." Let it be noted that this term is less than fully transparent. Allowing the related notions of "concept," "conception," and "conceivability" generous latitude, and extending this leniency to the term "cognition," the issue simply stated is: are there modes of cognition beyond human conception?

I do not believe that this issue is clearly stated, and I am doubtful that it can be much clarified. The situation is somewhat analogous to us, given our present knowledge and scientific-technological capabilities, asking why the Big Bang happened at the precise moment that it did, on the assumption that it was not immediately preceded by a Big Crunch. Our cosmological understanding is presently limited to events following the Big Bang. Similarly, allowing notions such as limits of conceivability, trying to conceive what might be beyond those limits is a prima facie absurdity.

It might be responded that beyond the limits of conceivability lies nonsense, and not simply nonsense for humans, but rather ontologically unqualified nonsense. Perhaps. But perhaps there is a third way between what is utterly nonsense and what is within our present and projectable cognitive domain. I realize that speaking of such a "third way" allows an unchallengeable escape from all criticism, and hence verges on vacuity. Only time will tell.

I have not shown that humans, in their present biological form, could not advance their knowledge to the point that they transcend what I am positing as limitations. When comparing ourselves to our momentarily supposed human descendants one hundred millennia hence, we might be in the position of, say, neolithic homo sapiens suddenly confronted with nuclear magnetic resonance imaging. I don't doubt the possibility of a cognitive-emphasized *human* existence in the deep future for reasons of inherent human cognitive limitations. There is little chance of predicting what humans would know, were humanity to survive and prosper for the next hundred thousand to one million years. It is rather that, as I have been urging, humans have other limitations imposed upon them by the demands of their successful evolution, and humans have possibilities they will realize that will transform them into posthumans (provided, of course, they do not first self-destruct), and these realized possibilities will include enhanced cognitive abilities that, I believe, are beyond possibility for biological humans.

In pursuing cognitive activity to what I take to be its limits with respect to cosmological study, and related studies of physics, mathematics, and logic, I have made speculative assumptions regarding the overcoming of these present limits. But, of course, with respect to cognition various presently conceived limits might be ultimate and unchangeable limits. Further, any attempt to transcend these limits by means of radical experimentation, such as has been suggested above, might have unforeseen, disastrous consequences.

Limits of cosmological, physical, mathematical, and logical explorations are not necessarily the relevant limits of other forms of knowledge. An example of another sort of knowledge limitation directly concerned with the overall theme of this book is our knowledge of the human brain. There is much known about the brain, and much that is unknown. Even as our understanding of biology overcame some previous limitations with the discovery of the structure, constitution, and functions of DNA, so will our understanding of brain functioning be largely enhanced if there are breakthroughs in understanding various details of coding, memory storage, and messaging within the brain.

On the one hand, the realization of such understanding of the human brain, or other brains, seems not such a deep future possibility so much as a future-to-far-future possibility. The brain is a biophysical organ, and what seems needed are mostly instruments with finer focus, able to trace individual neural events within the brain, as well as specifically delineated neuronal bundle activity. The devil might be in the details, and the details might yield to scientific and technological innovations in the not-so-distant future. On the other hand, the very complexity of the brain's operations might be beyond comprehension. The number of dendritic neural connections, actively growing and re-organizing throughout a human's lifetime, is in the trillions. The number of possible operations on such connections is astronomical, even under the simplifying assumption of two-value (on-off) activations of each connection.

This last difficulty is not irremediable. Consider the historical antecedent of the adoption of statistical methods of tracking the behavior of gases, largely due to the work of Boltzmann and Maxwell. There are an overwhelming number of atomic-molecular particles in any macro-volume of gas, each particle obeying classical mechanistic principles in any situation of chemical inertness. Yet the overall thermodynamic behavior of the system is predictable by these statistical methods, which also serve to explain that behavior at a deeper level than the explanatory patterns using the notion of gas as an elastic fluid.

In a similar manner, the intricate and multitudinous neuronal activity within the brain might someday be well summarized by somehow similar statistical methods. If such understanding can be achieved, it will likely happen in the near future. Hence, the question arises as to why such a scientific advancement is being discussed in this chapter. In response, the comparison to Boltzmann's breakthrough development of statistical mechanics is misleading, and the manner of disanalogy, once uncovered, provides an introduction to the sort of deep future possibility that might profoundly revolutionize whatever sentient cognition is operative at that time.

The methods of statistical mechanics allow a satisfying understanding of the thermodynamic behavior of particle states. Of course, the notion of satisfaction here is vague and intuitive, but in the context, and by comparison with what follows, it suffices. In contrast, from our current standpoint we would hardly be satisfied to know in significant detail the dynamics of the brain as some sort of physical system. Such enhanced understanding of brain activity would be of significant value both in itself and with regard to related physiological and psychological matters. But such understanding would not, by itself, substitute for our higher-level approaches to human behavior. Therein lies the disanalogy with statistical mechanics. There are well-known difficulties in the reductionist strategy of correlating brain states with specific aspects of these higher-level approaches.

Generally, the problem of correlating brain states with higher-level psychological states such as belief, desire, hatred, doubt, and so forth consists in what might be termed the "one-many" nature of the correlation. It is likely that different configurations of brain states (and, perhaps, future brain-surrogate states) are correlated with one sort of higher-level psychological state. In whatever manner brain states are typified, it is unlikely that such typification will pair with higher-level psychological typifications. This point has been made by such prominent philosophers of mind as Jerry Fodor[2] and Daniel Dennett, and can be found in the earlier writings of Hilary Putnam.[3] Consider, for example, the following passage from Dennett:

> Just as there are indefinitely many ways of *being a spark plug*—and one has not understood what an internal combustion engine is unless one realizes that a variety of different devices can be screwed into these sockets without affecting the performance of the engine—so there are indefinitely many ways of *ordering 500 shares of General Motors*, and there are societal sockets in which one of these ways will produce just about the same effect as any other. There are also societal pivot points, as it were, where which way people go depends on whether they *believe that p*, or *desire A*, and does not depend on any of the other infinitely many ways they may be alike or different.[4]

In context, Dennett is considering a deeper sort of reductionism than psychology as neurophysiology. Rather, he considers a Laplacean reduction of psychology to particle mechanics. The point of his discussion is to legitimize a quasi-realistic regard of emergent higher-level patterns. In Dennett's case, these patterns are the properties and relations of what he terms "intentional systems."

There has been much discussion of the various facets of this species of the reduction versus emergence issue, and it does not serve my purpose here to pursue the debate further. Referenced to current human knowledge and cognitive capability, my sympathies are largely with the emergentists. However, the deep future might realize possibilities that are not fully comprehendible by us.

If it were in any manner possible, would it serve humans to become thoroughgoing micro reductionists? But how would it be possible? Let us imagine that humans somehow come to experience the material world as a teeming mass of micro particle activity. There are roughly outlined areas of comparative rarefaction, as against comparatively denser areas. The relatively sharp density changes correspond to what actual humans perceive as edges of medium-sized objects. In this thought-experiment, the imagined humans do not consider themselves as humans in a world of tables, chairs, plants and animals. Nor do they consider themselves as thinking (or considering). They regard the world in the manner of advanced particle physicists, or perhaps as advanced wave theorists, or perhaps as both, or as something related, but other than these choices (strings?). Further, they experience the world in these terms. Tastes, colors, sensations; in short, the qualitative phenomena often held secondary to a natural scientific set of characteristics, are not present for them. Further still, they possess the cognitive ability to describe, explain, and navigate this world in realtime.

Two immediate questions arise. How can biological humans, present or future, do this, and, why would humans want to live in this manner? I take it that both these questions receive negative answers. Biological humans cannot and will not be able to experience the world like this. Humans would not want to experience the world in this manner. But, to the second question, what if sometime future humans could occasionally, on demand, experience the world in this manner? Of course, more needs be said here. What would be the duration of such experience? Roughly dividing the experience into cognitive and perceptual aspects, to what extent are either of these the result of activity internal to the human, and to what extent is the human relying on external apparatus?

I do not think these further questions need be addressed, on the assumption that they pertain to biological humans. It seems obvious that any biological humans, even if assisted by whatever apparatus can be developed, would be incapable of realtime processing of the sort of information presented to them in this speculation. However, our imagined deep future descendants (or, possibly, unitary descendant) might have capabilities enabling such processing. Seeking their reasons for wanting to engage in such activity, given that their human ancestors are incapable of it, is roughly analogous to asking why humans might want to study mathematics, given that gorillas are incapable of this study. This is not to demean human abilities as much as to acknowledge both their limitations, and that members of a species wants are often aligned with such limitations. We largely cannot imagine experiencing the world as the ontological microreductionist deems it to be because we are unable so to experience it, such experience would have scant realtime practical value for us, and we are by evolution accustomed to the experience of the world of medium-sized objects and processes.

We have obvious interest in the micro world. Our understanding of the micro world has had longstanding effect on our understanding of macro phenomena. Whatever our ontological position on the general issue of reduction versus emergence, however, it is the macro world we inhabit in our everyday lives. This need not be so for our speculated descendants. They might spend considerable time inhabiting the micro world, getting along in realtime experience of it. They might perceive this world by means internal to themselves, or by the aid of external instruments, although this external-internal distinction may be less than clearly delineated in their case, even as it is increasingly becoming blurred in our own case due to various perceptual aids and prostheses.

The manner in which these deep future posthumans conceive the micro world is worth a moment's speculation. They might be in possession of theories allowing for successful statistical treatment, akin to such statistical treatment as exists today. They might have cognitive abilities, perhaps with the aid of external devices, enabling realtime application of such statistical treatment. They might have, in addition or instead, theories and methods currently unknown to us. Given the time frame of one hundred thousand to one million years hence, this is not a bold supposition. Even as advances in mathematics astonished and invigorated our ancestors, so continuing theoretical and methodological advances would likely astonish us, were we able to witness them.

Most scientific advances in modern history have involved mathematics. Some of these advances have involved the creation of new mathematical forms, such as analytic geometry and calculus. Mathematical innovations will likely continue into the human and posthuman future. It may not be currently possible to conceive a nonmathematical replacement for mathematical theory and method. Such a replacement, if both conceivable and actual, would in some satisfactory manner accomplish tasks currently requiring mathematics, and surpass the range of possible mathematical applications. Our inability to conceive such a replacement notwithstanding, this supposition ought not to suffer peremptory rejection.

Of course, the "satisfactory manner" in which this hypothesized replacement for mathematics accomplishes and transcends what was accomplished by mathematics warrants further explication. While I haven't much to say regarding this, I want to note the following. Whatever this replacement might amount to, if it amounts to anything, the speculated fact that it is itself nonmathematical does not exclude mathematical criteria, such as consistency and precision, being applied to its results. Further, it does not exclude mathematical application of these criteria to its nonmathematical results. This replacement might in some recognizable sense be elegant, comprehensive, unifying, and workable, as is mathematics itself.

There are possible indications of what this replacement (or replacements) might resemble. Fractal geometries, cellular automata, and artificial neural networks are examples of techniques that have expanded our understanding of various physical domains. These, and other techniques, tend to serve as gestalts—organizing principles—allowing a particular perspective on physical states of affairs. Of course, the preceding three examples are not nonmathematical, but they are additions to what might be considered the higher level of the mathematical repertoire. That is, in their purely mathematical expression, these three techniques can be summarized in manners that do not indicate their physical manifestations, which are in effect interpretations of the formal symbolizations. It is the gestalt-suggestive generalizations of these interpretations that constitute the meaning of "higher level" in this context.

It would be misguided, I believe, to suggest a replacement of mathematics for its own sake. As I have indicated, I have no positive idea what such a replacement would be. I have merely gestured towards the possibility of such a replacement. I also recognize the possibility that in the deep future mathematics remains a primary instrument in the understanding of physical reality. Even as such, however, I do not believe we are currently able to predict the sorts of mathematics that will be relied upon in this understanding. Whatever intellectual instruments are then in place, they may somehow allow for a realization of that form of the reductionist's vision that physical reality can be described, explained, and predicted in the small, and that such understanding not only enriches our understanding of the macro world, but even further allows for a significant amount of prediction of higher-level macro world phenomena. That this is currently impossible, and perhaps as some seem to be urging, inconceivable, does not seem to me a prohibition on what might be the case one hundred thousand years hence.

In the case of those micro entities constituting brains or brain surrogates there arises the traditional problem of fatalism. If accurate higher-level predictions can be made regarding future activity of the sentient being possessing a brain or brain surrogate, then that being is apparently in the difficult position of knowing what it will do prior to the doing. However, there are various manners in which such difficulty might be avoided. All such predictions of macro events from micro phenomena might be probabilistic, rather than deterministic. There might be great practical difficulty in making such predictions of self or others' higher-level behavior. There might be a ban of some type on whatever apparatus is necessary in the service of such prediction. Or, perhaps most radically, the on-hand ability to make such predictions might not have the sort of consequences (resignation, depression) that are considered when applied to current humans.

There has been much focus above on what is currently considered the physics of the large and the small. The suggestion has been that posthuman life in the deep future might be occupied with this physics at what are currently the limits of the conceivable. If this seems a stultified existence, let me add that posthuman life might be at least as rich and variegated as current human life, and there might be a self-understanding of this life such as we cannot currently comprehend. There might further be a backward-looking understanding of our current human life, so as to offer currently inconceivable explanations of human affairs. I am suggesting that these speculated deep future descendants might not lack other modes of thought and experience than the cognitive. They will likely have a deep "scientific" understanding of these other modes. But they might also pursue these other modes with non-scientific attitudes, on some occasions.

NOTES

1. A good source of readings in these areas is Hilary Putnam and Paul Benacerraf eds., *The Philosophy of Mathematics: Selected Readings* (Cambridge, UK: Cambridge University Press, 1983).
2. Jerry Fodor, "Special Sciences (or: The Disunity of Science as a Working Hypothesis)." *Synthese* 28 (1974): 97–115.
3. Hilary Putnam, "The Nature of Mental States" in Ned Block, ed., *Readings in the Philosophy of Psychology*, Vol. 1 (Cambridge, Mass.: Harvard University Press, 1980), 223–31.
4. Daniel Dennett, *The Intentional Stance* (Cambridge, Mass.: MIT Press, 1990) 26.

Chapter 7

Loose Ends and Final Thoughts

There is an intuitive level of socio-political problematics at which are located discussions of race, sex, age, the environment, treatment of animals, and related concerns. Many humans in developed sectors would like to see racism, sexism, agism, pollution, and cruelty to animals either eliminated, or strictly curtailed. Popular media often report of progress in these endeavors, often keep a sharp eye on egregious negative instances, and occasionally raise awareness of chronic circumstances.

There are other levels of socio-political concern and discussion. Wealth disparities, lack of access to education, unemployment, healthcare issues, regional and international hostilities, and trade conflicts, are likewise the focus of attention. There are yet other levels vexing those who wish for a better world and, of course, the items in these various levels are often related both within the level and related to items in other levels.

Taking into consideration all that has preceded this chapter, it will be expected that I have easy responses to some of the above. I have maintained that, in time, the posthuman society of my speculations will be sexless, and consequently its members will not be sexually attracted to each other, nor will they explicitly or implicitly differentiate each other on the basis of sex. As they will be essentially ageless, age-based discrimination will not exist.

Questions regarding racial differences are not as easily handled, however. To the extent that race is a biological phenomenon, it is expected that the degree of control of biology our posthuman descendants possess will be sufficient to exorcize whatever biological differences create difficulties—if any difficulties remain. I admit that this is a quick answer to a problem that has troubled humanity for many centuries. I invite the reader to imagine scenarios in which control of biology allows for all sorts of physiological modifications. I believe that with some effort the reader will imagine at least one scenario in which those racial differences marked by physiology are eliminated to the satisfaction of a significant majority of the entire population, if not to everyone. As an aid in this thought-experiment, allow up to one thousand years for such realization and assume the following: age and sexual differences have been eliminated, all have indefinitely extended lifespan, and all have a roughly equal standard of living and are well-educated, by whatever advanced standards of education then exist.

As matters of race are currently understood, race is not solely a biological phenomenon. It is additionally, perhaps in some instances exclusively, a socio-cultural matter. Some speak of race as a socially constructed category, rather than a biological category. These non-biological understandings of race do not allow for the easy response, predicated on a speculated deeper mastery of biology in the future. To the extent that race is a socio-cultural matter, the problems occasioned by racial differences might be best addressed in the broader context of the difficulties generally attending social and cultural differences.

In the second chapter of this book I emphasized the lack of social scientific theories held in similar regard to those of the natural sciences. Given that lack, what follows is speculative, even with regard to human society, to say nothing of posthuman possibilities.

If much friction within a society can be immediately or eventually found to be the fault of economic disparities, then there is some likelihood that the amelioration of those disparities will bring significant lessening of social tensions. But perhaps this is too simple a suggestion. We should inquire into the causes of such economic disparities. Otherwise, there is the risk that any social engineering aimed at a minimally rough equalizing of economic circumstances may result in a temporary solution lacking long-term stability. Further, sizeable segments of various societies have shown resistance to programs whose planned outcomes include such equalizing.

I believe a deeper resolution of conflicts arising from chronic economic inequalities requires a basic alteration of human nature. I am speaking of the tendency of humans to outdo one another. Plato noted this tendency in his *Republic*, especially the first two books of that work, and therein created an imagined society in which that human characteristic would be expunged, albeit with only temporary success, amongst the guardians and rulers. In actual historical practice, such mitigation, let alone banishment, has not been effected for any appreciable period in any appreciable population. One could argue that on the whole humanity is better served by this characteristic; it has led to innovation and progress, which outweigh the evils it has wrought. Perhaps this has been the case in the earlier history of homo sapiens, up to the Industrial Age. It is questionable whether the human species is still better served by its members largely possessing this characteristic. This is obviously a complex question. Random individuals are likely not better served (by at least several intuitive standards of being so served) by the lack of this characteristic amidst a population overwhelmingly possessing it. Nor, given certain socio-economic and political circumstances, is a contemporary population as a whole better served by the absence of this characteristic. Nor, given the current abilities of the social and behavioral sciences, can confident projections be made about outcome for real populations regarding this question.

All the above granted, I want to hypothesize that, were the world population brought to an intuitively high economic and educational standard, humanity would face a better future absent this characteristic. I have struggled with the similar question of competition versus cooperation in previous chapters. I cannot adequately defend this hypothesis. But I can supplement it. Suppose that by some means humans were accorded the characteristic of having regard for the general welfare of humanity. Suppose further that they were given the characteristic of striving for such improvements as benefitted humanity (assuming some agreement as to such benefit), it being generally understood that competitiveness, or any characteristic fostering socio-economic inequality, is unacceptable. This allowed, is it not obvious that humanity would be in an improved condition, and that individual lives would be enriched?

Unfortunately for my intentions, I believe that many do not find this obvious. They have good reasons. There is the matter of specifying what benefits humanity, and justifying such specification. Absent a satisfactory account of such improvement (and a satisfactory account of what constitutes a satisfactory account), it is open to suggest that humanity is improved by engaging in the right sort of interpersonal competition. It will not settle matters for me to respond that this contrary suggestion is subject to identical criticism; namely, that the suggestion lacks specification and justification of its notion of improvement. Without further discussion, we are left with balanced alternatives.

I submit that there has been further discussion. Much that has preceded this chapter has been such further discussion in support of my proposal. I shall not review this further discussion here. Instead, I want to emphasize two points. We have been examining racial conflicts in particular and socio-cultural conflicts in general. I have been suggesting that if humanity is modified in ways I have indicated briefly here, and more thoroughly in previous chapters, such problems will have been dissolved.

One manner in which to solve socio-cultural conflicts, at least in one's imagination, is to equalize everyone's social status, and unify everyone's socio-cultural affiliations and proclivities. This is Plato's goal for the guardians of his proposed ideal state. Many find this repugnant. Some will see in this an ultimate realization of Nietzsche's notion of slave morality. Others will be reminded of Marx's criticisms of what he termed "crude communism:" ". . . the abstract negation of the entire world of culture and civilization . . ."[1]

Insofar as Nietzsche and Marx are speaking of a "leveling down," the modifications I have in mind do not suffer these criticisms. Put briefly, I seem to be considering a "leveling up." I accept the "upwards" part of the preceding brief characterization, but not the "leveling" half. Whatever comes to constitute culture, including cultural varieties, I do not envision the sort of homogenization that would fully answer to the charge of leveling. Even as today the various cultures of the developed societies are typically appreciated by those in the educated classes of any one of those societies, so I expect it to be in the speculated foreseeable future. I imagine our descendants enjoying the variations attendant their distinct future groupings. Unlike present cultural groupings, which typically fall along geopolitical and ethnic lines, future groupings might more easily form along lines of interest. Those with particular interests might be geographically dispersed, but I imagine them closely linked by communicative means exceeding those presently available.

I also imagine cultural groupings to be less a matter of ethnicity or nationality (indeed, if nationality remains at all), and more a matter of shared intellectual and aesthetic interests. Differences that initiate and sustain hostilities will hopefully be minimized. This hope is wishful to the extent that a natural-scientific sort of knowledge regarding such differences, together with the means to control these differences, is lacking. Possession of both such knowledge and capability of control render the mitigation of hostilities among groups a political matter. Mitigation of hostilities among groups has always been a political matter, but the political circumstances have heretofore been conjoined with various sorts of ignorance. The question posed here is: on the assumption that ignorance is removed, what remains?

More generally, the question is: what will politics be like in the future that is being considered here? Granted, this question cannot be answered with any degree of confidence. Nevertheless, I am supposing that the right sort of knowledge and technology—including biological and psychological knowledge and technology, will dramatically alter political relations and political activity. More particularly, political differences will be lessened to the point of disappearance of violent disagreements. It will be objected by some that there will always be political differences as long as there are dissensions. I agree that absent seemingly impossible changes in the foreseeable future, there will be conflicts of interest and other disagreements. What I am hoping is that overall circumstances will have changed to the point that our descendants will be other-regarding to a significantly greater extent than humanity has ever been, and that there will be a noteworthy second-order agreement regarding effective and peaceful methods of resolving conflicts.

Some might feel that the hope expressed in the previous paragraph "that overall circumstances will have changed" is utopian, in that people will never achieve such widespread and thorough peaceful coexistence. Or, if they do, it will be the result of intellect-dulling techniques such as are portrayed in well-known literary dystopias. I share this belief. I do not think human beings are capable of sustained peaceful coexistence, either among members of proximate groups, or on a global scale. I have given reasons for this in previous chapters, especially in "The Future of Ethics." Characteristics which aided human survival from neolithic homo sapiens to the present prevent such peaceful coexistence. I do not believe these characteristics can be removed without altering biological and psychological human nature to the degree that what results is no longer human. In welcoming the removal or significant lessening of these characteristics, I am welcoming posthumanity.

Even as not everyone values cooperation over competition, and many believe in the benefits of interpersonal competition, and many believe in retribution, so will many likely view devaluation of the human in favor of the posthuman, as has been done here, with disfavor. As I have given my reasons for humanity's transitioning in previous chapters, I want to consider the matter here, briefly, from a different perspective.

For all that I have said, both in this work and my previous work, I recognize that I might be grievously mistaken. By some measure that I do not believe we have yet discovered, but which would perhaps be universally accepted in the manner of a well-confirmed natural-scientific theory, it might transpire that humanity's best course is to remain human, rather than undergo a controlled evolution to something other than human. Should humans come to a general realization of this measure, and should it warrant the choice of remaining human, I would expect humans to take such actions as are necessary to remain human. I want to make the ontologically realistic supposition that such a measure exists but has not yet been discovered. That is, I want to suppose that however nuanced and qualified, however probabilistic, there are undiscovered answers of the form if these are the relevant circumstances, humanity is better off remaining human, if those are the relevant circumstances, humanity is better off transitioning to posthumanity.

If it seems I have been giving an unqualified answer to the question of transitioning, allow me to correct that impression. I believe it likely that humanity will be better served by transitioning, but lacking the right sort of theory, and lacking a comprehensive sampling of the data constituting the domain of the theory, whatever sort that data is, I am ontologically, if not psychologically, humbled.

It would be preferable for humanity to be able to return to its human state, if having transitioned only to find posthumanity an untenable or barely tolerable mode of being. Without constructing a detailed speculative scenario, there are imaginable manners in which such return might not be possible. In which case, humanity will have made a disastrously wrong decision. It might be that, transitioned to posthumanity, nothing frightful results, but there is a general longing to return. Again, it would be preferable that, having experienced both, the general wish for return could be satisfied. And again, this wish might be unrealizable. Should return be possible, there will likely be attendant suffering, even as there will be suffering in the large-scale transition to posthumanity. Any movement involving great numbers of people has been accompanied with the various sufferings of dislocation. There is little reason to suppose a difference in this respect regarding the possibilities we are contemplating here.

Game-theoretic optimization strategies might suggest that, with the above taken into account, the favored option is to remain human. All other considered options appear to involve some amount of suffering that will not occur if we persevere in our humanity. This calculation would be secure provided two conditions are met. One, the time frame is kept relatively short. That is, we do not consider an extended length of time including both the transition to posthumanity with its attendant suffering, and an appreciable length of time following the transition in which perhaps posthumanity achieves a quality of life considerably greater than humanity enjoyed. This remark of course assumes an acceptable measure of quality of life, as does the earlier assumption of a barely tolerable posthuman existence.

The second condition that needs be met to ground the speculated game-theoretic judgment is that nothing intolerable occur to the general human population that would not have occurred had the transition been effected. I believe this is a problematic assumption. If humans remain human in the present biological understanding of being-human, each individual human will die. To some humans, mortality is intolerable. To those finding mortality intolerable, some simply live with the knowledge that they are mortal; that is, they tolerate the intolerable. Others either convince themselves, or believe on faith, that they will have an immortal existence after their present finite existence is done.

Let us suppose what is likely the case, that being mortal is not an intolerable condition to many humans. There are almost certainly some humans for whom the overall extinction of humanity is not an intolerable circumstance (I am speaking here of extinction *tout court*, and not replacement by posthumanity). There might be some humans for whom devolution to post-nuclear-apocalypse scavengers amidst the rubble is a fond thought. These sorts of cases aside, I believe that for the majority of humans the overall extinction of humanity is an intolerable thought. Yet if we remain as humans for a bit longer there is some incalculable probability that this will be the fate of the species, long before it is the physical fate of our solar system. There are reasons for this assessment which are sufficiently well-known as not to need listing here. It is possible, perhaps probable, that controlled evolution to something other than human is a better option than remaining human.

Speculating about our posthuman future is entertaining to some. Very few, however, have a sense of urgency regarding the matters discussed in this book. I have no such sense of urgency. On the other hand, many have concerns, local and global, about the present state of humanity. Once again, very few find their concerns regarding humanity mollified by thoughts of transitioning to posthumanity. Because I believe such transitioning is not an immediate (i.e., within the next decade or two) prospect, and because I believe that if and when it commences, the transitioning will be accompanied by a significant amount of suffering (as I stated in the chapter titled "On Transition"), I am among this group of those not regarding our posthuman future as a present-day panacea for present ills.

I do not know of a present-day panacea for present ills. It is questionable whether there is any listing of present ills that would have near-worldwide agreement. Perhaps there are several items on nearly every thoughtful person's list. I would place hunger, adequately defined, on such a list. Yet there are those who are apparently convinced that the world is overpopulated, and that hunger is one manifestation of this circumstance. As such, present-day hunger is understood to be symptomatic of present ills, rather than a fundamental ill itself. While I disagree with this neo-Malthusian position, I recognize that any intellectually proper settling of this disagreement requires a careful assessment of various factual matters, somc of which are in dispute.

A listing of present ills, such as was done at the beginning of this chapter, is an exercise in pessimism. And, one can pursue a pessimistic listing of ills that go beyond present ills, such as future shortages. Of course, a listing of workable correctives for these ills is an exercise in optimism. Then there is a further optimistic list of possibilities which follow on the assumption that such correctives have been successfully instantiated. That is a good deal of what I have been about in this book, although as I discuss in the chapter "On Transition," I doubt that the transition to posthumanity that I anticipate will be neatly founded on a prior alleviation of most of society's ills.

Rather than continue a listing of present ills, I would like to suggest social means of mitigation of those ills. It is understood that employment of any large-scale social means is in part a political matter. The importance of political considerations noted, I have nothing to suggest here regarding the political means to achieve these social means. I don't have adequate beliefs as to what such political means would be.

It is further understood that a sufficient amount of material wealth must be available to implement various social means, including those which I shall propose. I simply assume that adequate wealth exists, and that its expenditure in service of the social means I recommend is largely a political matter.

As I have already suggested in "The Future of Ethics," humanity is incapable of realizing an ultimately satisfactory state of affairs (allowing tolerance of the phrase "ultimately satisfactory"). I hold this due to deficiencies in human nature, such circumstances being deficiencies when measured against what I take to be best possibilities for humanity, if only these possibilities were realizable. In other words, our best possibilities as humans are ideal possibilities, to be realized, if at all, by posthumans. Hence, my suggestion of social means of mitigation of present ills should be understood as partial and short term.

Given that my suggestion is woefully general, disappointing, and controversial, I first present and support the fundamental premise of the suggestion, after which the suggestion itself will be swiftly stated.

Many years ago I came to the position that ethical disagreements are largely, if not fully, factual disagreements. I came to this position as a result of my teaching introductory ethics courses, in the context of discussions of ethical relativism. Many of my students would cite seemingly obvious differences in ethical judgments, often based on different cultural and historical circumstances. To provoke further thought, I would counter that historical differences regarding astronomical beliefs do not render the facts of these astronomical matters relative to the varied believers. In short, I challenged them to show the relevant dissimilarity between ethical matters and matters of natural science.

There were students who had little difficulty with this challenge. They extended their ontological reach to the formulation of a distinction between mere opinion, as against opinion capable in principle of factual resolution. For some time I was stymied, until I had (or read, or heard; I forget) an idea familiar to others: what if the great majority of humanity does not have ethical disagreements? What follows if it is a fact of human nature that humans largely and fundamentally share values? Suppose ethical disagreements are mostly apparent. Then it becomes a major task of ethics to discover that with which we are largely and fundamentally in agreement. In this task, practitioners of ethics are joined by those practitioners of social and natural sciences whose work is of relevance to the task.

I took as a model example the case of the committed Nazi, by whom I did not mean the mere functionary or apparatchik. Rather, I intended the person committed to Nazi ideology, who in the course of duties sanctions the murder of Jews, Gypsies, the mentally afflicted; who sanctions performing hideous experiments on Poles; who sanctions the enslavement of Slavs; the person who does this fully believing these acts are morally justified. Surely this person, of whom there were many, had different values than the rest of us.

Except, this person believed that the only fully human beings are Aryans, and that Slavs, Gypsies, and so on are sub-human, somehow either less than fully human, or of a lesser order of human than the Aryan. Even as most humans believe that nonhuman animals are not accorded the full rights that humans are granted, and that the higher up the phylogenetic scale nonhuman creatures are positioned (typically, largely in terms of their cognitive abilities), the more consideration they ought be accorded. Put more succinctly, the more they are like us, the more they are to be treated like us.

The committed Nazi does not view those condemned as human, or as fully human. Further, the committed Nazi sees those condemned as a threat of one sort or another to those who are fully human. They might interbreed with Aryans, thereby producing defective offspring. They sometimes insinuate themselves into the Aryan social structure, thereby weakening it either with lesser ideas, or subversive activities. And so forth.

Hopefully, it is recognized that the committed Nazi holds factually mistaken beliefs. How this sort of person comes to acquire these beliefs, and what might be involved in ridding such people of these beliefs, is a matter about which I have no insights to offer. But, unlike the case of those Nazis who went along to get along, I wonder at the hypothetical result of such ideologically committed people being disabused of these beliefs. At one time, the Nazi believes what has been ascribed to the Nazi's system of beliefs. Imagine a short time later that the formerly committed Nazi comes to believe the negation of these previous beliefs. Clearly, what will then happen is a psychological matter. Various defense mechanisms might be activated. Where before there was unwavering commitment, now there is perhaps wavering rationalizations. Perhaps now the formerly ideologically committed Nazi is satisfied to adopt the role of mere functionary, satisfied to survive in difficult times. It is difficult to predict how this will go. But I believe one thing is a near certainty. The formerly ideologically committed Nazi will not have the full moral conviction that he possessed previously, unless he has suffered a severe psychological breakdown.

I have discussed this model case at some length. Yet I am aware of the deficiencies within it. There are vague notions and hand-waving inferences. Even granting the possibility that these sorts of deficiencies could be rectified, there remains the outstanding difficulty of showing that this model case is representative of all putatively fundamental value disagreements. Needless to say, I cannot overcome these difficulties here. Perhaps they cannot be overcome. Perhaps not all fundamental moral disagreements are, in principle, resolvable factual disagreements. Perhaps various significant factual disagreements are fundamentally normative disagreements, rather than the converse I am maintaining. I believe normative disagreements are fundamentally factual disagreements, and I have hinted by example at the outlines of a more general defense of my belief, but I must leave it at that. I shall proceed with my suggestion based upon the premise, the possibly mistaken premise, that fundamental disagreements in value are reducible to factual disagreements.

From this premise I draw the suggestion that we humans are obligated to learn and widely disseminate the relevant facts. I believe this learning and dissemination will be accompanied by action appropriate to these facts. If, for example, it is shown to be a fact that all our behavior is fully governed by natural laws, and this fact is understood by a world wide population of humans educated to standards that are themselves factually based as high standards, I expect that retributive punishment will be seen as contrary to the facts, and will no longer be considered a justification of our treatment of any individual.

The relevant facts regarding value disagreements will be seen to be of various sorts. Among the most immediately relevant facts will be those facts regarding whatever wants are most fundamental among humans. I believe a fundamental want most people have, at most moments of their lives, is that their lives should continue, rather than terminate at that moment. I believe most people want to live flourishing lives, and that further understanding of various facts will show that people largely agree regarding what constitutes such lives.

There is, of course, much more to be said here. There are reasons to doubt and to deny what I have been saying, vague and general though it has been. I should add that I do not think the sort of facts that are relevant to this enterprise are socially constructed, or in any decisive manner relative to the mode of inquiry by which we are made aware of them. I know that many will object to this metaphysical attitude. I will give them the benefit of agreeing that they might be correct about this, even by my standards of correctness. Because I do not think they are correct, I suggest that humanity's best option is to create a world in which most of us are capable of learning and implementing the relevant and objective facts of living well.

To come full circle: I see the above suggestion as the best humanity can manage in the short term. If I am correct about this, and if humanity manages it well, then what I have predicted regarding the difficult transitioning of humanity to posthumanity will not be as troublesome as I have supposed. I would like humanity to know significantly more than it currently does before individuals begin to transition. Were that so, those gathering on the other side will have better ideas of what next to do.

NOTE

1. Karl Marx, *The Economic and Philosophical Manuscripts of 1844*, Martin Mulligan, trans. (New York: International Publishers, 1964) 133–134.

Selected Bibliography

Aristotle. *Nichomachean Ethics*, trans. Martin Oswald. New York: Bobbs-Merrill, 1962.

Austin, John. *Sense and Sensibilia*. New York: Oxford University Press, 1962.

Bostrom, Nick. "A History of Transhumanist Thought." *Journal of Evolution and Technology* 14, no.1 (April 2005): 1–25.

———. "Why I Want to Be Posthuman When I Grow Up." in *Medical Enhancement and Posthumanity*, eds. Bert Gordijn and Ruth Chadwick, 107–37. Dordrecht: Springer, 2008.

———. "The Future of Humanity." in *New Waves of Philosophy in Technology*, eds. Jan-Kyrre Olsen and Evan Selinger, 186–215. New York: Palgrave Macmillan, 2009.

Bostrom, Nick, and Julian Savulescu, eds. *Human Enhancement*. New York: Oxford University Press, 2009.

Chalmers, David. *The Conscious Mind: In Search of a Fundamental Theory*. New York: Oxford University Press, 1996.

Cooney, Brian. *Posthumanity: Thinking Philosophically about the Future*. Lanham, Md.: Rowman & Littlefield, 2004.

De Grey, Aubrey, with Michael Rae. *Ending Aging: The Rejuvenation Breakthroughs That Could Reverse Human Aging in Our Lifetime*. New York: St. Martin's Press, 2007.

Dennett, Daniel. *Elbow Room: The Varieties of Free Will Worth Wanting*. Cambridge, Mass.: MIT Press, 1984.

———. *The Intentional Stance*. Cambridge, Mass.: MIT Press, 1990.

———. *Consciousness Explained*. Boston: Little, Brown and Company,1991.

Descartes, Rene. *Meditations on First Philosophy*, trans. Elizabeth Haldane and G. Ross, *The Philosophical Works of Descartes* Vol. I. New York: Dover, 1931.

Dostoevski, Fyodor. *The Brothers Karamazov*, trans. Constance Garnett. New York: Random House, 1995.

Eddington, Arthur. *The Nature of the Physical World*. New York: Macmillan, 1928.

Faulkner, William. *The Sound and the Fury*. New York: Garland, 1986.

Fischer, John. *The Metaphysics of Free Will: An Essay on Control*. Cambridge, Mass.: Blackwell,1995.

Fodor, Jerry. "Special Sciences (or: The Disunity of Science as a Working Hyposthesis)." *Synthese* 28, no. 2 (October, 1974): 97–115.

Freud, Sigmund. *Civilization and Its Discontents*. New York: J. Cape and H. Smith, 1930.

Fukuyama, Francis. *Our Posthuman Future: Consequences of the Biotechnology Revolution*. New York: Farrar, Straus and Giroux, 2002.

Garreau, Joel. *Radical Evolution: The Promise and Peril of Enhancing Our Minds, Our Bodies—and What It Means to Be Human*. New York: Doubleday, 2005.

Hughes, James. *Citizen Cyborg: Why Democratic Societies Must Respond To The Redesigned Human Of The Future*. Cambridge, Mass.: Westview Press, 2004.

Hume, David. *Dialogues Concerning Natural Religion*. Indiannapolis, Ind.: Hackett, 1980.

Kant, Immanuel. *Critique of Pure Reason*, trans. Norman Kemp Smith. London: Macmillan, 1963.

———. *Critique of Practical Reason*, trans. Lewis White Beck. New York: Macmillan, 1993.

Krauss, Lawrence. *The Physics of Star Trek*. New York: HarperCollins,1995.

Kuhn, Thomas. *The Structure of Scientific Revolutions*. 2d ed. Chicago: University of Chicago Press, 1970.

Libet, Benjamin. "Do We Have Free Will?" *Journal of Consciousness Studies* 6, nos. 8–9 (1999): 47–57.

Marx, Karl. *The Economic and Philosophical Manuscripts of 1844*, trans. Martin Mulligan. New York: International Publishers, 1964.

———. *A Contribution to the Critique of Political Economy*. New York: International Publishers, 1970.

Mill, John Stuart. *Utilitarianism*. New York: Bobbs-Merrill, 1957.

Naam, Ramez. *More Than Human: Embracing the Promise of Biological Enhancement*. New York: Broadway Books, 2005.

Nietzsche, Friedrich. *The Portable Nietzsche*, trans. and ed. Walter Kaufmann. New York: Viking, 1954.

Parfit, Derek. *Reasons and Persons*. New York: Oxford University Press, 1984.

Pike, Nelson. *God and Evil*. Englewood Cliffs, New Jersey: Prentice-Hall, 1964.

Plato, *Gorgias*, trans. Donald Zeyl. Indianapolis: Hackett, 1987.

———. *Phaedo*, trans. G. M. A. Grube. Indianapolis: Hackett, 1977.

———. *Protagoras*, trans. Stanley Lombardo and Karen Bell. Indianapolis: Hackett, 1992.

———. *Republic*, trans. G. M. A. Grube and C. D. C. Reeve. Indianapolis: Hackett, 1992

———. *Theaetetus*, trans. M. J. Levett and Myles Burnyeat. Indianapolis: Hackett, 1992.

Putnam, Hilary. "The Nature of Mental States." In Ned Block, ed., *Readings in the Philosophy of Psychology*, Vol. 1, 223–31. Cambridge, Mass.: Harvard University Press, 1980.

Putnam, Hilary and Paul Benacerraf eds., *The Philosophy of Mathematics: Selected Readings*. Cambridge, UK: Cambridge University Press, 1983.

Sartre, Jean-Paul. *Existentialism and Human Emotions*, trans. Bernard Frechtman and Hzel Barnes. New York: Citadel Press, 1957.

Schulz-Aellen, Marie-Françoise. *Aging and Human Longevity*. Boston: Birkhäuser, 1997.

Seidel, Asher. *Inhuman Thoughts: Philosophical Explorations of Posthumanity*. Lanham, Md.: Lexington Books, 2008.

Shostak, Stanley. *Becoming Immortal: Combining Cloning and Stem-Cell Therapy*. Albany: State University of New York Press, 2002.

Skinner, Burhuss Frederic. *Walden Two*. New York: Macmillan, 1976.

Spinoza, Baruch. *The Ethics and Selected Letters*, trans. Samuel Shirley. Indianapolis: Hackett, 1992.

Stanovich, Keith. *The Robot's Rebellion*. Chicago: University of Chicago Press, 2004.

Strawson, Peter Frederick. *Freedom and Resentment and Other Essays*. London: Methuen, 1974.

Turing, Alan. "Computing Machinery and Intelligence." *Mind* 59, no. 236 (1950): 433–60.

Van Inwagen, Peter. *An Essay on Free Will*. New York: Oxford University Press, 1983.

Young, Simon. *Designer Evolution: A Transhumanist Manifesto*. Amherst, N.Y.: Prometheus, 2006.

Index of Proper Names

www.ingramcontent.com/pod-product-compliance
Lightning Source LLC
LaVergne TN
LVHW101448200926
848483LV00001B/93

* 9 7 8 0 7 3 9 1 5 0 0 6 1 *